HARCOURT SCHOOL PUBL

Think Math!

Texas Edition

Student Work Text

Lesson Activity Book

Developed by Education Development Center, Inc. through National Science Foundation

Grant No. ESI-0099093

EDC

Harcourt
SCHOOL PUBLISHERS

Visit *The Learning Site!*
www.harcourtschool.com/thinkmath

Printed in the United States of America

ISBN 13: 978-0-15-358853-2

ISBN 10: 0-15-358853-5

2 3 4 5 6 7 8 9 10 048 16 15 14 13 12 11 10 09 08

This program was funded in part through the National Science Foundation under Grant No. ESI-0099093. Any opinions, findings, and conclusions or recommendations expressed in this program are those of the authors and do not necessarily reflect the views of the National Science Foundation.

Principal Investigator

E. Paul Goldenberg

Curriculum Design and Pedagogy Oversight

| E. Paul Goldenberg | Lynn Goldsmith | Nina Shteingold |

Research

Director: Lynn Goldsmith	Sabita Chopra	Suenita Lawrence
Nina Arshavsky	Sophia Cohen	Katherine Schwinden
Cynthia Char	Andrea Humez	Eugenia Steingold

Editorial

| **Director: Frances Fanning** | Nicholas Bozard | Eric Karnowski |

Writing

Director: Eric Karnowski

Jean Benson	Stacy Grossman	Paisley Rossetti
Abigail Branch	Andrea Humez	Nina Shteingold
Sara Cremer	Suenita Lawrence	Kate Snow
E. Paul Goldenberg	Debora Rosenfeld	Julie Zeringue

Graphics and Design

Directors: Laura Koval and Korynn Kirchwey

| Jessica Cummings | E. Charles Snow |
| Jennifer Putnam | Jenny Wong |

Project Management

Directors: Eric Karnowski and Glenn Natali

Amy Borowko	Alexander Kirchwey	Kimberly Newson
Nannette Feurzeig	Helen Lebowitz	David O'Neil
Kim Foster	June Mark	Cynthia Plouff

Mathematics Reviewers

Richard Askey, Professor of Mathematics, Emeritus
University of Wisconsin, Madison, Wisconsin

Harvey Keynes, Professor of Mathematics
University of Minnesota, Minneapolis, Minnesota

Roger Howe, Professor of Mathematics
Yale University, New Haven, Connecticut

David Singer, Professor of Mathematics
Case Western Reserve University, Cleveland, Ohio

Sherman Stein, Professor of Mathematics, Emeritus
University of California at Davis, Davis, California

Additional Mathematics Resource

Al Cuoco, Center Director, Center for Mathematics Education, Education Development Center, Newton, Massachusetts

Advisors

Peter Braunfeld	June Mark
David Carraher	Ricardo Nemirovsky
Carole Greenes	James Newton
Claire Groden	Judith Roitman
Deborah Schifter	

Evaluators

| Douglas H. Clements | Mark Jenness |
| Cynthia Halderson | Julie Sarama |

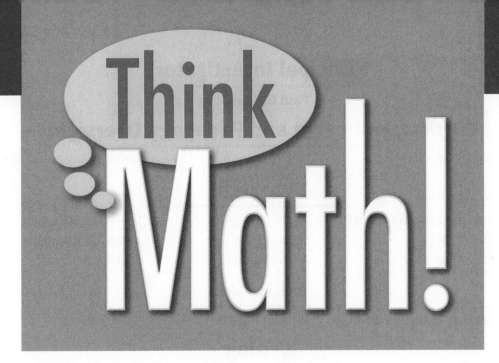

Think Math!

Chapter 1 Two-Dimensional Figures and Patterns

Chapter 2   Number Lines and Time

Think Math! Contents

Chapter 5　Working with Tens

Chapter 6　Data and Probability

Contents

Chapter 7   Working with Larger Numbers

Chapter 8   Doubling, Halving, and Fractions

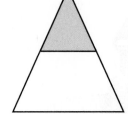
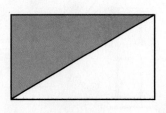
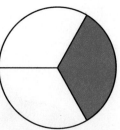

Chapter 9   Modeling Addition and Subtraction

Chapter 10   Maps, Grids, and Geometric Figures

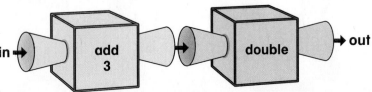

Think Math! *Contents*

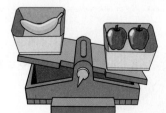

Chapter 13　Making and Breaking Numbers

Chapter 14　Extending Addition and Subtraction

Contents

Chapter 15 Exploring Rules and Patterns

Chapter 1

Two-Dimensional Figures and Patterns
Comparing and Sorting ✏️
🔺 **TEKS 1.6C**

Sort the objects into two groups.

STEP 1 **Showing the Groups**

Draw or trace the objects to show how you sorted. Write a name for each group.

_____   _____

STEP 2 **Sorting Another Way**

Sort the objects another way.
Tell about it.

_____

_____

_____

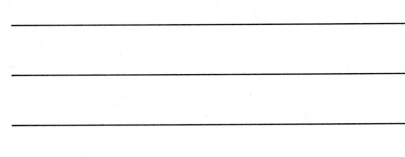

Investigation

Dear Family,

Today we started Chapter I of *Think Math!* In this chapter I will learn about two-dimensional figures and how they are similar and how they are different. I will also learn about patterns and ways to recognize and make patterns with objects, figures, and numbers. There are NOTES on some of my pages to explain what I am learning every day.

Here are some activities for us to do together at home. These activities will help me understand two-dimensional figures and patterns.

Love,

_____

Family Fun

Shape Dominoes

Play this game with your child to provide practice drawing and recognizing two-dimensional figures. Your child will play a similar game in school.

■ You will need a red crayon, a blue crayon, and a yellow crayon.

■ On a sheet of paper, draw a small or large triangle, circle, square or rectangle using one of the crayons.

■ Have your child draw a figure next to yours that is different in only one way—size, shape, or color.

■ Switch roles and repeat. Play until you have created a long chain of figures. Have your child tell how the figures differ each time.

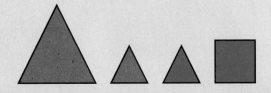

Household Patterns

Work with your child to create patterns with common household objects.

■ Use common household objects such as forks and spoons, napkins, coins, or paper clips to create patterns.

■ Arrange the objects in a row to create a repeating pattern by alternating the object in a certain way, or by placing similar objects in different positions.  For example, you might make a pattern using 2 forks, I spoon, 2 forks, I spoon, or by placing paper clips in different positions.

■ Repeat the activity with different objects. Encourage your child to describe each pattern.

Introducing *Think Math!*

NCTM Standards 1, 6, 7, 8, 9, 10
TEKS 1.1D, 1.5A

I. Write the number or draw the picture.

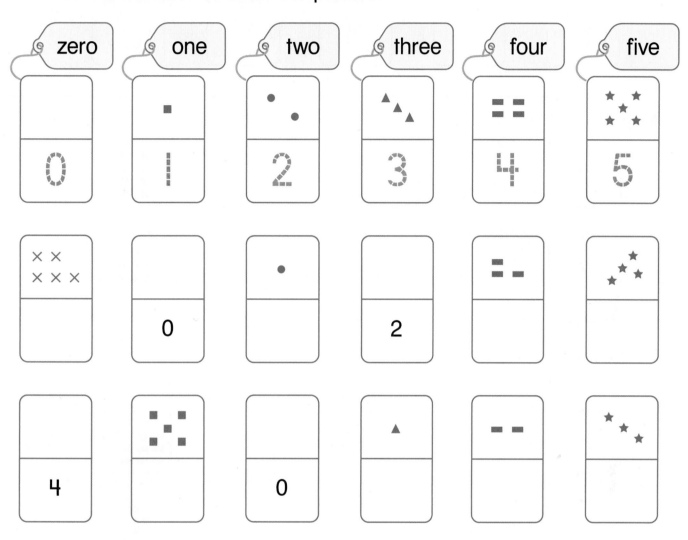

2. Draw each set.

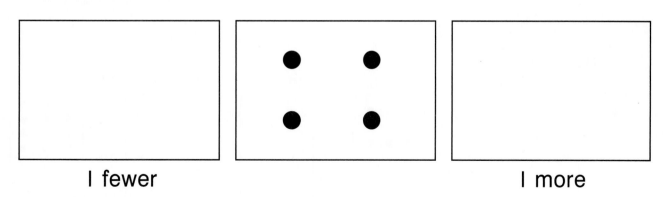

I fewer   I more

NOTE: Your child is learning to write numbers to match sets of objects. Together, find patterns within the rows and columns on this page.

     III   three   3

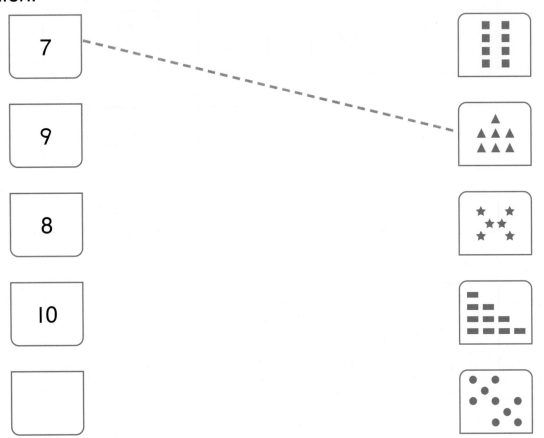

six   seven   eight   nine   ten

6   7   8   9   10

3. Match.

7

9

8

10

[ ]

Problem Solving

4. What is missing?

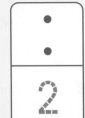 2    4   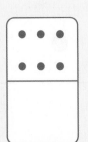       10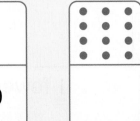

Name _____ Date _____

Examining Two-Dimensional Figures

NCTM Standards 3, 6, 7, 8, 9, 10
🔻 **TEKS 1.6A, 1.13**

1. Match.

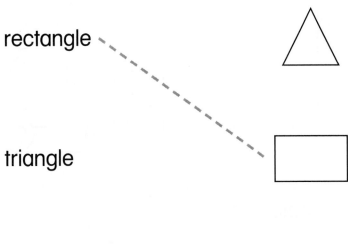

rectangle

triangle

circle

---

2. Color ▭ blue.

Color △ green.

Color ○ red.

🏠 **NOTE:** Your child is learning to identify and describe two-dimensional figures. Ask your child to name the figures on this page.

I ✋   🪙 **V** five ▊ 5

3. Color 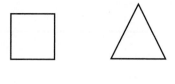. Write how many.

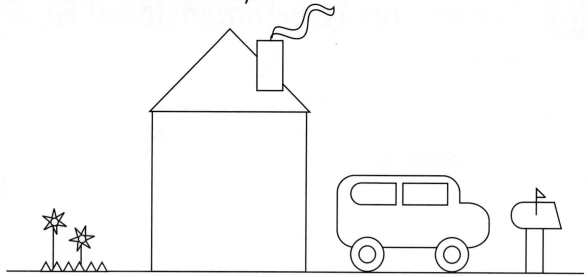

There are _____ rectangles.

Which figure does not belong? Cross it out.

4.

5.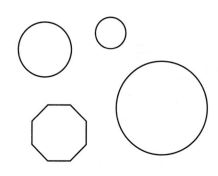

Challenge

6. Which figure does not belong?
Draw a figure that does belong.

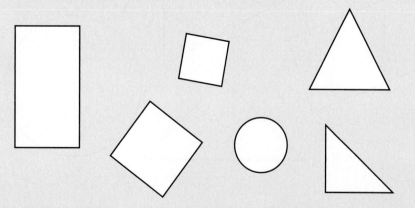

Name _____ Date _____

Sorting by Attributes

NCTM Standards 3, 5, 6, 7, 8, 9, 10

🔺 TEKS 1.6C, 1.13

Which does not belong? Cross it out.

1.

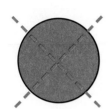 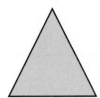

2.

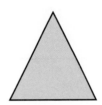

3.

4.

5.

6.

NOTE: Your child is learning to sort objects by color, shape, and size. If you like, ask your child to describe how two figures on this page are similar and how they are different.

  VII seven **7**

Draw another figure that belongs.
Explain why.

7.

_____

8.

  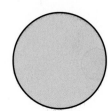

_____

9.

_____

Challenge

10. The figure with the X does not belong. Why?

 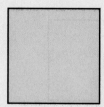

_____

8 eight **VIII**

Sorting by More Than One Attribute

NCTM Standards 3, 6, 8, 9, 10

🔻 **TEKS 1.6C, 1.13**

What is missing?

1.

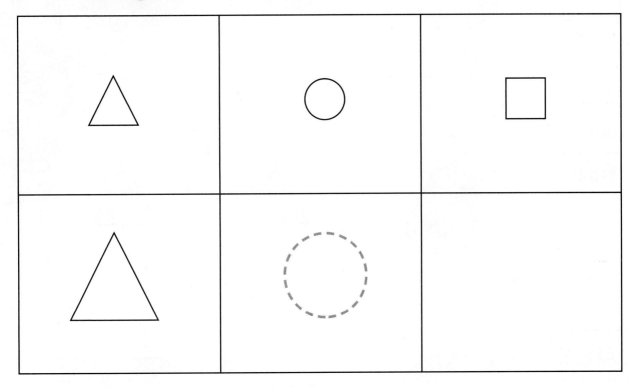

2.

 NOTE: Your child is learning to identify similarities and differences. Have your child tell how the figures on this page are the same and how they are different.

   IX nine 9

Which figure is it?

3.

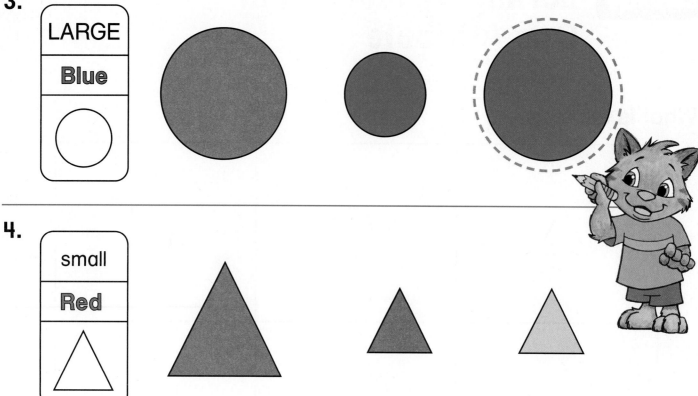

4.

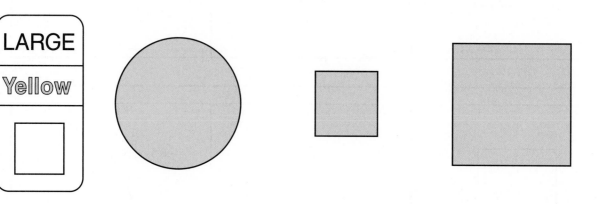

5.

Challenge

6.

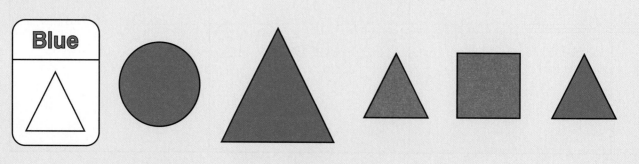

 10 ten **X**   **2**

Name _____ Date _____

Counting Differences

NCTM Standards 1, 3, 6, 7, 9, 10
🌵 TEKS 1.6C, 1.13

Draw figures that are different in one way each time.

1.

2.

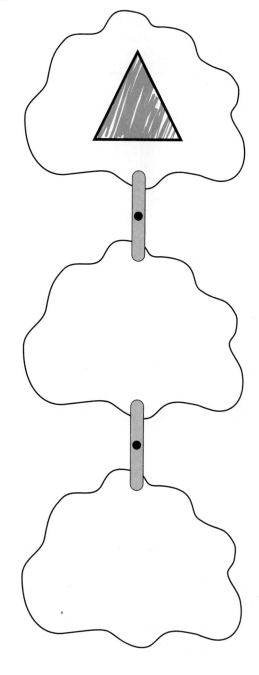

NOTE: Your child is learning to count the number of ways things can be different. These figures can differ in shape, size, and color.

   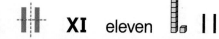

3. Draw figures that are different in two ways.

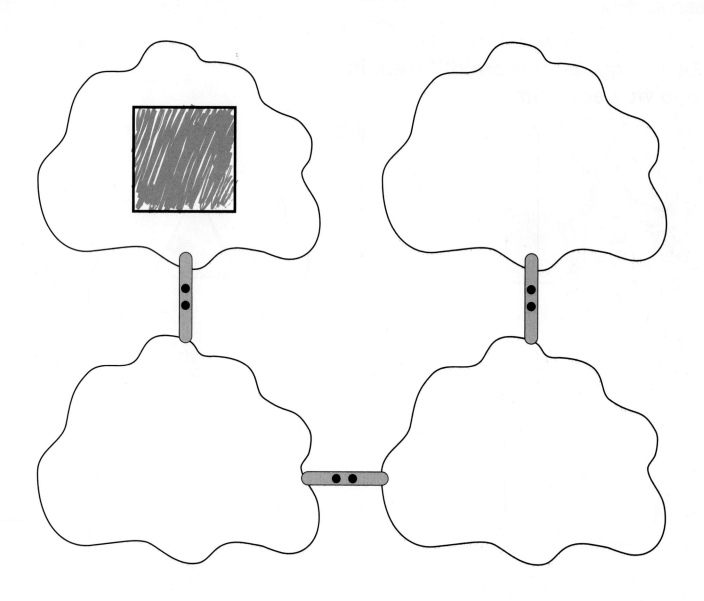

Challenge

4. In how many ways are the figures different?
Draw the dots.

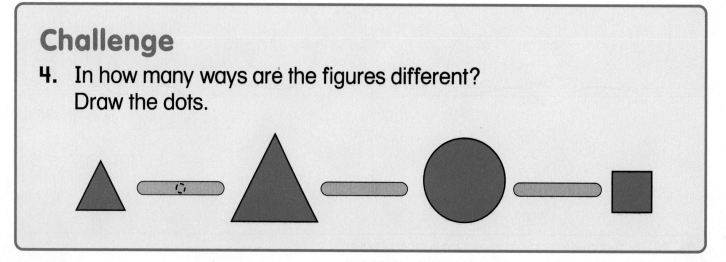

Review/Assessment

NCTM Standards 2, 3, 6, 7, 8, 9, 10

Write how many. Lesson 1

1.

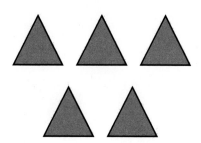

_____

2.

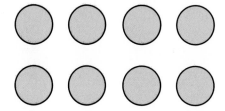

_____

Cross out the figure that does not belong. Lessons 2 and 3

3.

  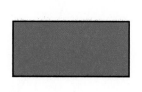 

4.

5.

Draw each figure. _{Lesson 4}

6. large   red   square

7. small   blue   circle

8. Draw a figure that is different in one way. _{Lesson 5}

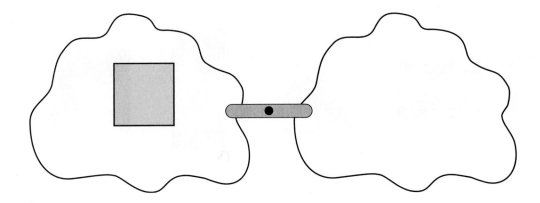

Problem Solving _{Lessons 6 and 7}

What comes next?

9. ▲ ▲ ● ▲ ▲ ● ▲ ▲ ____

10. ☐ ☐☐ ☐☐☐ ☐☐☐ _____

18 ‖ eighteen **XVIII** 18 $9 + 9$

Chapter 2

Number Lines and Time
Lining Up Cubes ✏️
📍 TEKS 1.1A

You need
- 20 connecting cubes

**Work with a partner.
Each makes a cube train.**

STEP 1 Drawing the Cube Trains

Your Train

Your Partner's Train

STEP 2 Comparing the Trains

Which train has more cubes? How do you know?

_____

_____

_____

STEP 3 Making a Bigger Train

How many cubes do you need to make a bigger train? Explain.

_____

_____

_____

Investigation

Dear Family,

Today we started Chapter 2 of *Think Math!* In this chapter, I will learn about number lines and see that jumping forward on a number line is adding, and jumping back is subtracting. I will also explore time and learn to tell time to the hour. There are NOTES on the Lesson Activity Book pages to explain what I am learning every day.

Here are some activities for us to do together at home. These activities will help me learn to add, subtract, and tell time.

Love,

_____

Family Fun

What's My Number?

Work with your child to identify a mystery number.

■ One player thinks of a number from 0 to 10.

■ The other player asks *yes/no* questions to discover the mystery number. Here are some sample questions

"Is the number less than 5?"

"Is the number greater than 2?"

"Is the number 4?"

■ If there is a number that cannot be the mystery number, you may wish to draw a number line and cover the number with a dry bean or other small object.

■ Try to find the mystery number with the fewest questions.

0 1 2 3 4 5 6 7 8 9 10

Time Tic-Tac-Toe

Work with your child to investigate time to the hour.

■ Copy this gameboard or make up a similar one of your own.

7:00 Morning	9:00 Morning	6:00 Morning
3:00 Afternoon	1:00 Afternoon	2:00 Afternoon
5:00 Evening	6:00 Evening	7:00 Evening

■ The first player chooses a time on the board and thinks of an activity that is done at that time. If both players agree that the time for that activity is reasonable, that player places an "X" on that time.

■ The second player names an activity and if both players agree, places an "O" on that time.

■ The first player to mark three squares in a row, column, or diagonal, wins.

Introducing the Number Line

NCTM Standards 1, 2, 6, 9, 10

🔺 TEKS 1.5B

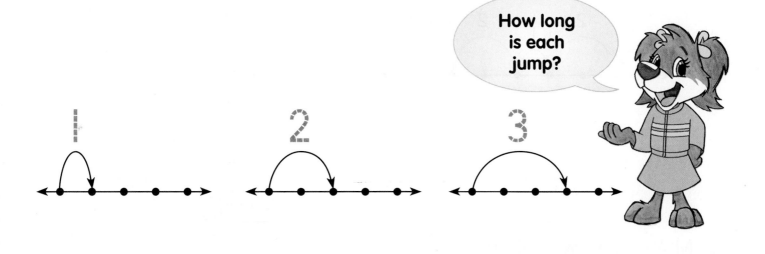

Continue each pattern.

1.

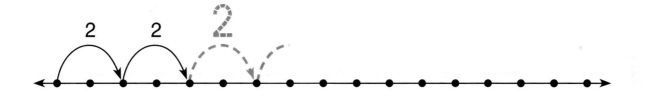

2.

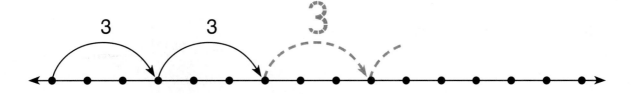

🏠 **NOTE:** Your child is beginning to learn about number lines.
In this lesson, children draw jumps on a dot line and
continue patterns.

20 + 1   🔺 **XXI**   twenty-one   21

3. Continue the pattern.

---

4. Make your own pattern.

Challenge

5. Continue the pattern.

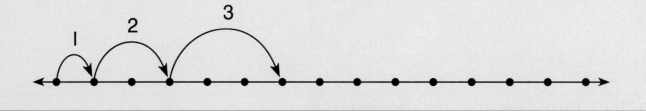

Jumping on the Number Line

NCTM Standards 1, 2, 6, 8, 9, 10

🔶 TEKS 1.5B

What is missing?

1.

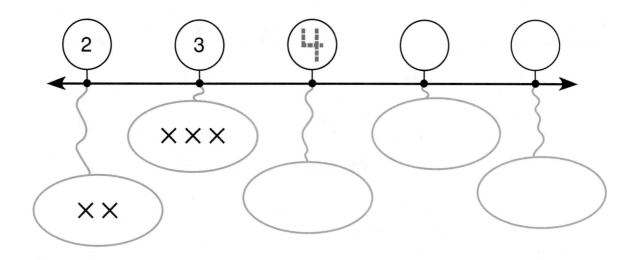

2.

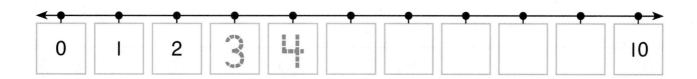

3. Continue the pattern.

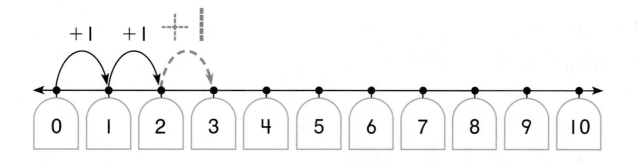

© Education Development Center, Inc.

🏠 **NOTE:** Your child is beginning to identify missing
numbers on a number line and to see that a jump
forward can be indicated with a + sign,
and a jump backward with a − sign.

Continue each pattern.
Describe the jump.

4.

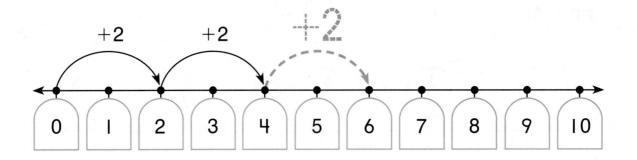

Jump forward _____ spaces.

5.

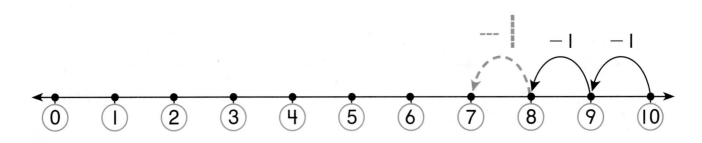

Jump backward _____ space.

Challenge

6. Continue the pattern.

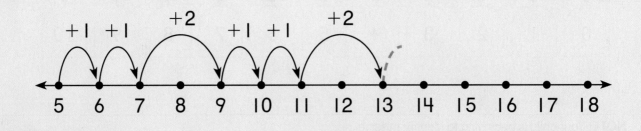

Recording Jumps as Addition and Subtraction

NCTM Standards 1, 2, 6, 7, 9, 10

Find the missing jumps and numbers.

1.

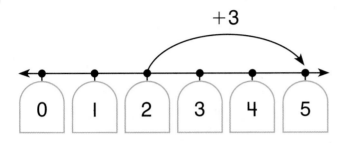

2.

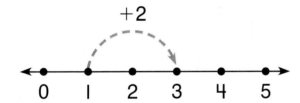

3.

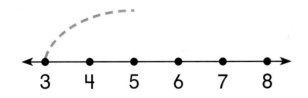

NOTE: Your child is learning to connect addition and subtraction with jumps on a number line. A jump forward is addition. A jump backward is subtraction.

Find the missing numbers.

4.

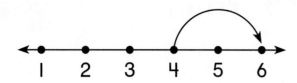

$$\begin{array}{|c|} \hline 4 \\ \hline \end{array}$$
$$\bigcirc$$
$$\begin{array}{|c|} \hline 6 \\ \hline \end{array}$$

5.

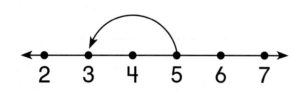

$$\begin{array}{|c|} \hline 5 \\ \hline \end{array}$$
$$\boxed{-\ 2}$$
$$\begin{array}{|c|} \hline \\ \hline \end{array}$$

Problem Solving

6. Draw the jump.
Find the missing number.

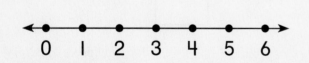

$$\begin{array}{|c|} \hline \\ \hline \end{array}$$
$$\boxed{+\ 4}$$
$$\begin{array}{|c|} \hline 6 \\ \hline \end{array}$$

Name _____ Date _____

Relating Addition and Subtraction

NCTM Standards 1, 2, 6, 9, 10

 TEKS 1.3B

Draw the jump. Find the missing numbers.

1.

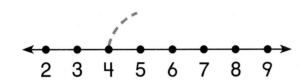

2 3 4 5 6 7 8 9

| 4 |
| (+ 3) |
| 7 |

2.

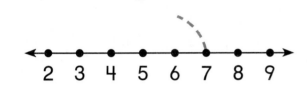

2 3 4 5 6 7 8 9

| 7 |
| (− 3) |
| |

3.

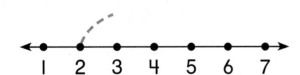

1 2 3 4 5 6 7

| 2 |
| |
| 6 |

4.

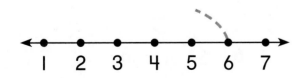

1 2 3 4 5 6 7

| 6 |
| |
| 2 |

NOTE: Your child is learning that addition and subtraction *undo* each other. If you jump forward 3 spaces on the number line then jump backward 3 spaces, you will end up where you started.

Draw the jump. Find the missing numbers.

5.

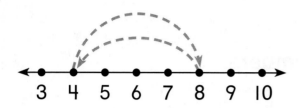

4	8
+ 4	− 4

6.

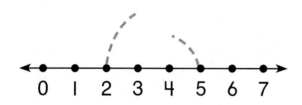

2	5
+ 3	− 3

7.

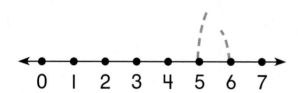

5	6
◯	◯
6	5

Challenge

8.

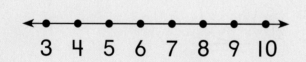

	9
+ 6	− 6
9	

Comparing Numbers on the Number Line

NCTM Standards 1, 2, 6, 7, 8, 9, 10

TEKS 1.1A

Write the missing number or symbol.

1.

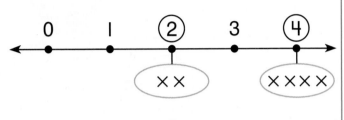

2 (<) 4

2.

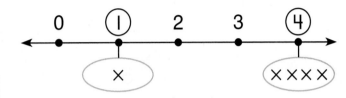

1 ( ) 4

3.

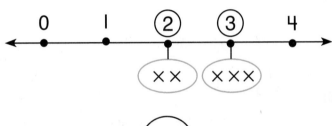

3 ( ) 2

4.

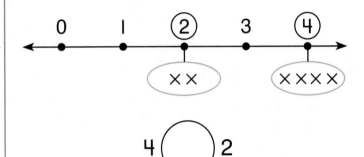

4 ( ) 2

5.

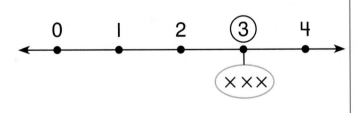

3 = ☐

6.

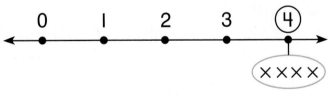

4 > ☐

NOTE: Your child is learning to use symbols to compare numbers and sets of objects. Ask your child to tell you what the symbols >, <, and = mean.

7. What is missing?

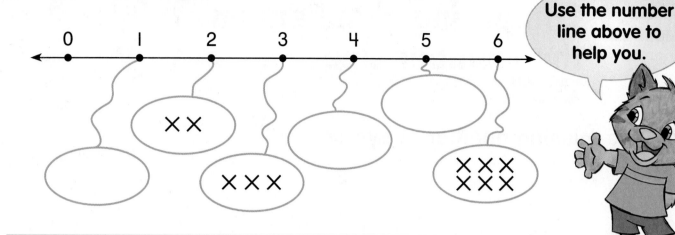

Use the number line above to help you.

8.	2 $<$ 5		**9.**	1 $\bigcirc$ 6
10.	4 $\bigcirc$ 3		**11.**	2 $\bigcirc$ 2
12.	3 $\bigcirc$ 0		**13.**	1 $\bigcirc$ 5

 14. Look at the number line.
How can you tell that 5 is greater than 3?

_____

_____

Challenge

15. Write the numbers. Draw the Xs.

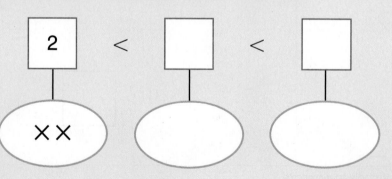

2 $<$ ☐ $<$ ☐

Name _____ Date _____

Comparing Numbers and Quantities

NCTM Standards 1, 2, 6, 7, 9, 10

TEKS 1.1A, 1.5B, 1.12A

Write <, >, or =.

1.

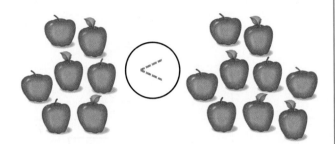

2.

3.

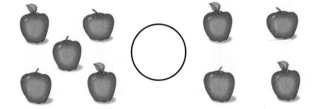

4.

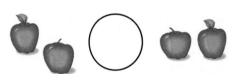

5.

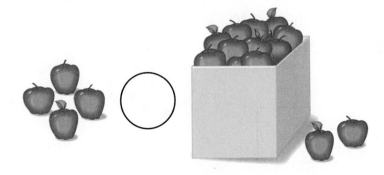

NOTE: Your child is learning to compare the number of objects in two sets. Ask your child to explain how to compare each pair without counting.

6. What is missing?

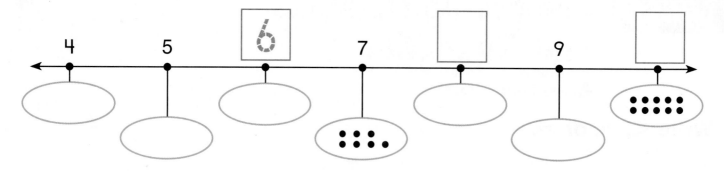

Compare.

7. 8 $\big(\!>\!\big)$ 7

8. 5 $\bigcirc$ 9

9. 6 $\bigcirc$ 4

---

10. ☐ $<$ 10

11. 10 $>$ ☐

12. 10 $=$ ☐

Make your own.

13. ☐ $>$ ☐

14. ☐ $<$ ☐

Problem Solving

15. Which creature has more legs? _____
Write a number sentence to
compare the numbers.

☐ $\bigcirc$ ☐

spider   **ladybug**

Chapter 2
Lesson 7

Investigating Time and Events

NCTM Standards 4, 6, 7, 8, 9, 10

TEKS 1.8A

I. Draw something you do **before** school.

2. Draw something you do **after** school.

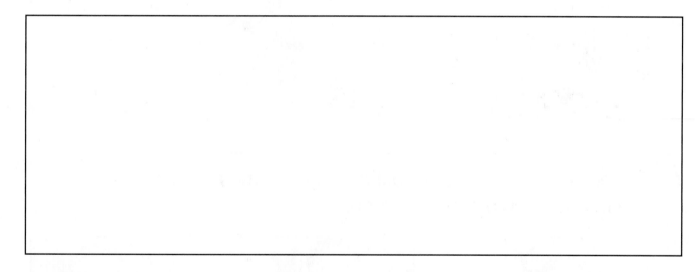

3. Which takes more time? Explain.

_____

_____

_____

Work with a partner. Do one activity.
Your partner does the other.
Start at the same time.
Circle the activity that takes longer.

4. Write your name.

Count to 30 by ones.

5. Clap 10 times.

Jump up and down 10 times.

6. Order the activities from least amount of time to most.
Write 1, 2, or 3 to show the order.

Eat breakfast _____   Be in school _____   Brush teeth _____

Chapter 2
Lesson 8

Telling Time to the Hour

NCTM Standards 1, 4, 6, 7, 9, 10

TEKS 1.8B

What time is it?

1.

2.

3.

4.

5.

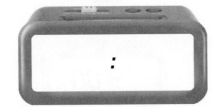

6.

NOTE: Your child is learning to tell time to the hour
and to distinguish the hour hand from the minute hand.

Draw the missing hour hand.

7.

11:00

8.

1:00

9.

7:00

10.

12:00

Challenge
Draw the missing hands.

11.

6:00

12.

3:00

thirty-six **XXXVI** 3 dozen

Ordinal Numbers and the Calendar

NCTM Standards 1, 6, 7, 8, 9, 10

Complete the calendar for this month.

_____

Sunday	Monday	Tuesday	Wednesday	Thursday	Friday	Saturday

1. Color the second Monday yellow.

2. Color the first Friday blue.

3. Color the tenth day of the month red.

4. What day is the fifth day of the month? _____

5. What is the ninth day of the month? _____

The boy below is getting ready for school.
Put the things he does in order.

first	second	third	fourth	fifth	sixth
1st	2nd	3rd	4th	5th	6th

first

_____

_____

_____

_____

_____

Problem Solving Strategy
Work Backward

NCTM Standards 1, 2, 4, 6, 7, 8, 9, 10

 TEKS 1.8B, 1.11B

Understand
Plan
Solve
Check

1. Cary plays outside for 2 hours.
He stops at 5 o'clock.
What time did Cary start playing?

```
┌─────────┐
│    :    │
└─────────┘
```

---

2. Abby had 7 baseball cards.
Now she has 9 cards.
How many more cards did she get?

_____ more cards

---

3. Steve and his sister have 10 minutes
until bedtime.
They play a game for 4 minutes.
How much more time do they have to play?

_____ minutes

NOTE: Your child is exploring different ways to solve
problems. Sometimes you need to use the information and
work backward to solve a problem.

Problem Solving Test Prep

1. Harry draws a figure with 4 sides and 4 corners.

What figure could he have drawn?

(A) triangle   (C) square

(B) heart   (D) circle

2. Ethan writes this pattern.

A   C   E   G   I   K

What letter comes next?

(A) L   (C) N

(B) M   (D) O

 Show What You Know

3. Kaylie makes this pattern.

What figure comes next?

_____

Explain how you know.

_____

_____

_____

_____

4. Jenny has 6 shells. She finds 1 more. How many shells does Jenny have now?

_____ shells

Explain.

_____

_____

I. **What numbers are missing?** Lessons 1 and 2

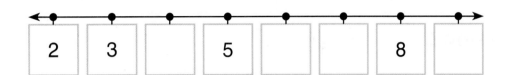

| 2 | 3 | | 5 | | | 8 | |

Find the missing numbers and jumps. Lessons 3 and 4

2.

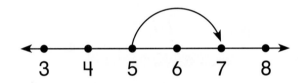

5

$+ 2$

☐

3.

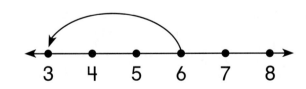

6

◯

3

Compare. Write >, <, or =. Lessons 5 and 6

4.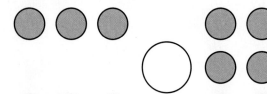

5.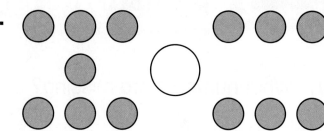

What is missing? Lesson 8

6.

7.

8. What day is the second day of the month? Lesson 9

_____

JANUARY

Sunday	Monday	Tuesday	Wednesday	Thursday	Friday	Saturday
			1	2	3	4
5	6	7	8	9	10	11

Problem Solving Lesson 10

9. It takes 1 hour for the bus to get to the city.
The bus gets in at 7:00.
What time did the bus leave?

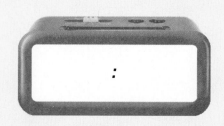

© Education Development Center, Inc.

Name _____

Skip-Counting and Money

Counting Pennies

Put a handful of pennies on your desk.

STEP 1 Guessing How Many

Do you have more or less than 10 pennies? _____

About how many pennies do you have?

about _____ pennies

STEP 2 Counting the Pennies

Count the pennies by ones.

How many pennies do you have?

_____ pennies

Was your guess close? _____

STEP 3 Counting Another Way

Tell another way to count the pennies.

_____

Which way do you like better? Why?

_____

_____

_____

Investigation

Dear Family,

Today we started Chapter 3 of *Think Math!* In this chapter, I will find the value of collections of pennies and nickels and make amounts of money in different ways. I will also practice skip-counting. There are NOTES on the *Lesson Activity Book* pages to explain what I am learning every day.

Here are some activities for us to do together at home. These activities will help me understand money and learn strategies for counting coins.

Love,

_____

Family Fun

Trading Game

Work with your child to practice trading 5 pennies for 1 nickel.

- Gather 40 pennies, 8 nickels, and a number cube.

- Players take turns tossing the number cube and taking that number of pennies.

- When a player has 5 pennies, he or she must trade the pennies for 1 nickel.

- Players find the value of the coins after each turn. Values can be recorded in a table.

Turn	Player 1	Player 2
1	4¢	6¢
2	7¢	8¢

- The first player to collect 20¢ wins.

Ways to Pay

Work with your child to practice making amounts using pennies and nickels.

- Use self-stick notes or pieces of paper to make price tags. Write a price less than 25¢ on each tag. Put the tags on small items such as pencils, fruit, and snacks.

- Have your child select one of the items to "buy." Give your child 5 nickels and 20 pennies. Ask your child to choose coins to pay for the item.

- Challenge your child to show another way to pay for the item using a different combination of coins.

- Repeat the activity with different items and different prices.

Introducing the Nickel

NCTM Standards 1, 2, 4, 6, 7, 8, 9, 10

TEKS 1.1C, 1.3A

Glue nickels and pennies to make different amounts. Then write how much.

Total

penny

nickel

	¢		¢		¢

1.

2.

3.

4.

NOTE: Your child is learning to add nickels and pennies to find the total amount. Show your child some nickels and pennies and ask your child to find the total amount.

50 – 1   **XLIX**   forty-nine   49

Write how much in the table.

5.

6.

7.

8.

9.

10.

	How Much?
5.	6 ____ ¢
6.	____ ¢
7.	____ ¢
8.	____ ¢
9.	____ ¢
10.	____ ¢

Challenge

Write <, =, or >.

2 < 3	2 = 2	2 > 1

11.  ◯

12.  ◯

50 | fifty **L** | 10

Counting Money

NCTM Standards 1, 2, 4, 6, 7, 8, 9, 10

TEKS 1.1C, 1.11A, 1.11B, 1.11C, 1.11D

How many coins are there?
What is the value?

 P is a penny, worth 1¢. 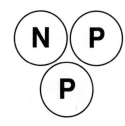 **N** is a nickel, worth 5¢.

1.

P

_____ coin

_____ ¢

2.

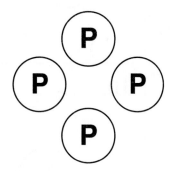

_____ coins

_____ ¢

3.

N **P**
P

_____ coins

_____ ¢

4.

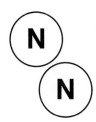

_____ coins

_____ ¢

5.

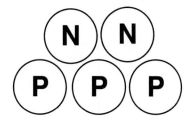

_____ coins

_____ ¢

6.

_____ coins

_____ ¢

NOTE: Your child is learning how to find the value of collections of nickels and pennies. If you like, have your child use coins to show different amounts to 10¢.

Write how much in the table.

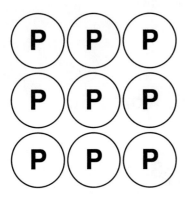 is a penny, worth 1¢. 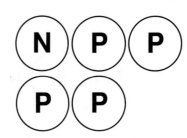 is a nickel, worth 5¢.

	How Much?
7.	_____ ¢
8.	_____ ¢
9.	_____ ¢
10.	_____ ¢

7.

P P P
P P P
P P P

8.

N P P
P P

9.

P P P P
P P P P
P P P P
P P P P

10.

N N
N P

11. Draw your own. Show 2 different ways to make the same amount.

Problem Solving

12. I have 18¢. I have some pennies and nickels. What coins could I have?

Chapter 3
Lesson 5

Making Amounts in Different Ways

NCTM Standards 1, 2, 4, 6, 7, 8, 10

TEKS 1.11A, 1.11B, 1.11C, 1.11D, 1.12A, 1.13

Draw ways to make 10¢.

Use Ⓝ for nickels and Ⓟ for pennies.

1.

2.

3.

 4. Did you show all of the ways to make 10¢?
Explain how you know.

_____

_____

_____

 NOTE: Your child is learning that there are different ways to make the same amount of money. If you like, ask your child to show different ways to make an amount using nickels and pennies.

Color to show the amount.

 is a nickel, worth 5¢. is a penny, worth 1¢.

5.

   8¢

6.

   15¢

7.

   19¢

Problem Solving

8. Use only nickels and pennies.
How many different ways can
you show 12¢? Show each way.

_____ ways

Chapter 3
Lesson 6

Problem Solving Strategy
Act It Out

NCTM Standards 1, 2, 4, 6, 7, 8, 9, 10

🔸 TEKS 1.11A, 1.11B

Understand
Plan
Solve
Check

I. William and Jenny have these coins.

William	**Jenny**

How much money do they have together?

_____¢

---

2. Kyle had 2 pennies.
Carlos gave him a nickel.
Tina gave him 4 pennies.
How much money does he have now?

_____¢

---

3. Mary had 14 cents.
She bought some gum for a nickel.
How much money does she have left?

_____¢

NOTE: Your child is exploring different ways to solve
problems. Acting out a problem can help children
understand what is happening in a problem.

Problem Solving Test Prep

1. Jason found 4 pennies.
Now he has 6 pennies.
How many did he start with?

(A) 2 pennies (C) 6 pennies

(B) 4 pennies (D) 10 pennies

2. Which number comes between 10 and 15?

(A) 9 (C) 13

(B) 10 (D) 16

 Show What You Know

3. Kira has a soccer game at 4:00. The book club meets at 3:00. She will eat dinner with her family at 6:00.

Which activity will Kira do first?

_____

Explain.

_____

_____

_____

_____

_____

_____

4. Cameron made a pattern.

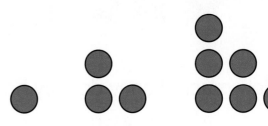

Draw what comes next.

Describe the pattern.

_____

_____

_____

Chapter **3**   **Review/Assessment**

NCTM Standards 1, 2, 4, 6, 7, 8, 9, 10

What is the value?   Lessons 1, 3, 4

1.

_____ ¢

2.

_____ ¢

3.

_____ ¢

4.

_____ ¢

How many beads are green?   Lesson 2

5.

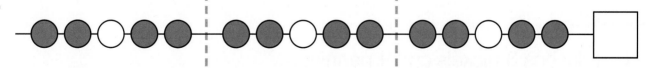

6.

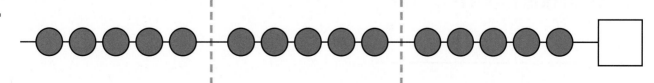

What is the total value? Lesson 3

7.

 = _____ ¢

8.

 = _____ ¢

Color to show 11¢ two ways. Lesson 5

9.

10.

Problem Solving Lesson 6

11. Billy has 3 nickels and 4 pennies.
Anita has 2 nickels and 8 pennies.
Who has more money?

_____

Name _____

Exploring Addition and Subtraction

Add in Any Order

Put some counters on each plate.

STEP 1 Counting

How many counters are on the left plate? _____

How many counters are on the right plate? _____

STEP 2 Finding the Total

How many counters are there in all? _____

Explain how you found the total.

_____

STEP 3 Adding in Another Order

Switch the plates so that the left one is on the right.

How many counters are there in all now? _____

How does this compare to the answer in Step 2?

_____

Why do you think the totals are the same?

_____

_____

Dear Family,

Today we started Chapter 4 of *Think Math!* In this chapter, I will use stories and puzzles to learn about addition and subtraction. There are NOTES on the Lesson Activity Book pages to explain what I am learning every day.

Here are some activities for us to do together at home. These activities will help me learn to add and subtract.

Love,

_____

Family Fun

Missing Number

Work with your child to identify a missing number of items.

- Two players need exactly 10 paper clips or other small objects for this game. Set the 10 objects on a table.

- One player hides some of the 10 objects in his or her hand and challenges the other player to tell how many objects are hidden.

- By counting the number of objects left, the player determines the number of hidden objects.

- If the number is correct, that player scores 1 point.

- Players take turns. The first player to get 10 points wins!

Tic-Tac-Add

Work with your child to add two numbers.

- Draw a tic-tac-toe board with the numbers 1, 2, 3, and 4 in the four squares of the top left corner.

- Assign your child to be X and to go first. Players take turns and follow the rules of tic-tac-toe with one modification: if a player wants to mark a square without a number, he or she must add the two numbers in the row, column, or diagonal and write the sum on the game board first before marking an X or O.

- Play continues until one player makes tic-tac-toe or the game ends in a tie. Players can play again by marking the upper left squares of a new board with the numbers 2, 3, 4, 5; 1, 1, 2, 4; or 2, 2, 4, 5 or any other combination of four numbers less than 5.

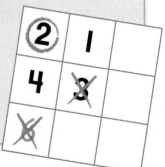

Introducing the Cross Number Puzzle

NCTM Standards 1, 2, 3, 6, 8, 9, 10

🔸 **TEKS 1.6C**

What numbers are missing?

1.

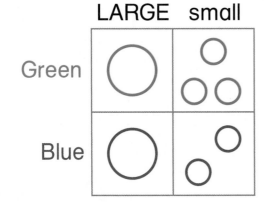

	LARGE	small
Green	1	3
Blue		

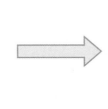

2.

	□	○
LARGE		
small		

3.

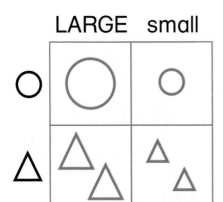

	LARGE	small
○		
△		

NOTE: Your child is learning to sort shapes, count them, and record the numbers in a table. This prepares children for exploring addition and subtraction in later lessons.

4. What numbers are missing?

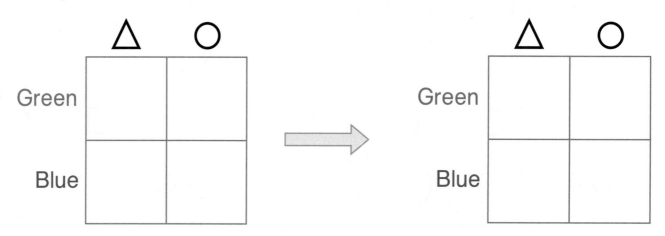

Green Blue

LARGE

small

→

Green Blue

LARGE

small

---

5. Make your own.

△ ○

Green

Blue

→

△ ○

Green

Blue

Challenge

6. What figures are missing?

○ □

small

LARGE

→

○ □

	○	□
small	3	4
LARGE	1	2

Name _____ Date _____

Using Cross Number Puzzles to Add

NCTM Standards 1, 2, 3, 6, 8, 9, 10

TEKS 1.3B, 1.12A

Complete each puzzle.

1.

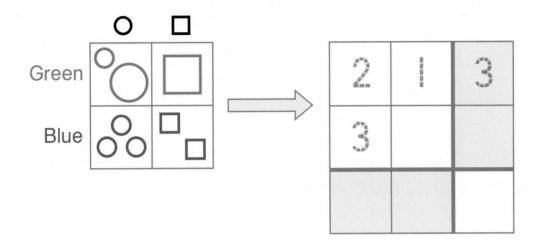

2.

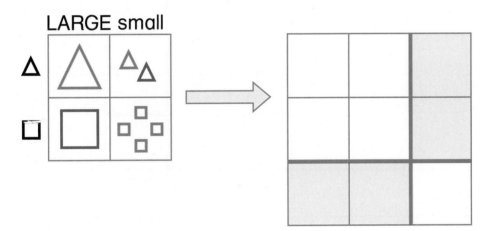

3.

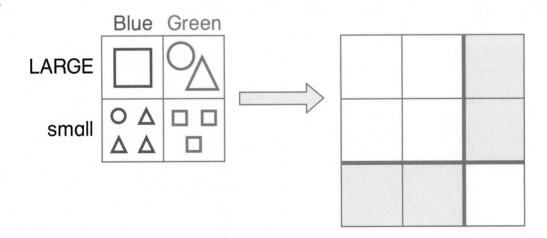

NOTE: Your child is learning to count objects in a grid and to record the numbers in a Cross Number Puzzle. In these puzzles, sums of objects in rows and columns are recorded in the bottom row and in the right column.

4. What are the missing numbers?

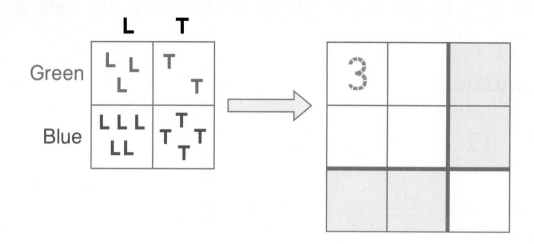

5. What are the missing letters and sums?

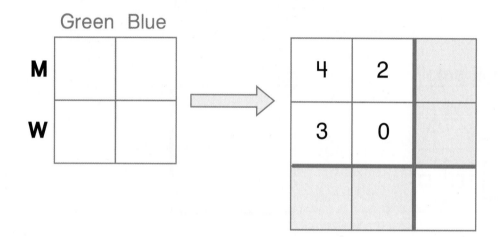

Challenge

6. What are the missing numbers and shapes?

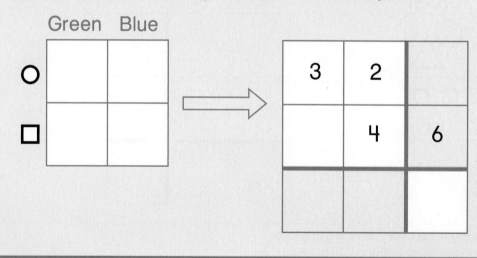

32 + 32

Exploring Missing Addends

NCTM Standards 1, 2, 6, 7, 8, 9, 10

🔺 TEKS 1.3B

Fill in the missing pictures and numbers.

> Use the numbers in the puzzles to decide how many to draw.

1.

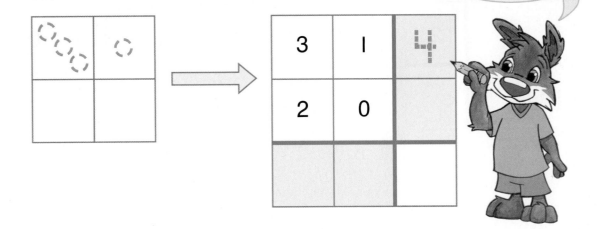

2.

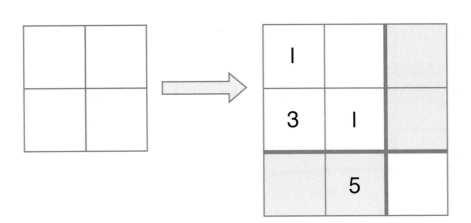

3.

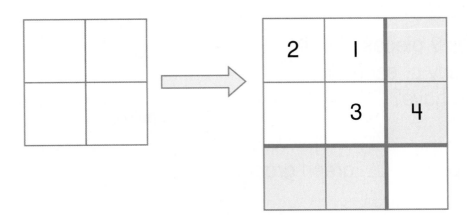

NOTE: Your child is using reasoning skills to complete Cross Number Puzzles. The numbers in the puzzle match the number of circles in the 2-by-2 grid. Ask your child to explain how to solve each puzzle.

What are the missing numbers?

4.

	2	3
0		2
1		

$\square$ + 2 = 3

0 + $\square$ = 2

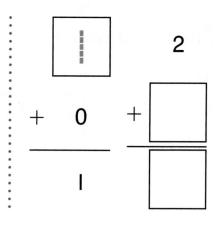

$\square$ 2

+ 0   + $\square$

   1   $\square$

5.

3		7
	1	
4		

3 + $\square$ = 7

$\square$ + 1 = $\square$

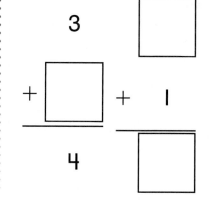

  3   $\square$

+ $\square$ + 1

  4   $\square$

Problem Solving

6. Trey has 9 pieces of fruit. How many of each kind does he have?

red grapes _____ green grapes _____

red apples _____ green apples _____

	red	green	
grapes			5
apples		1	
	3		9

⫾ + ⫾ + ⫾ + ⫾ + ⫾ + ⫾

Name ______________________ Date __________

Chapter 4
Lesson 4

Practice with Cross Number Puzzles

NCTM Standards 1, 2, 6, 7, 8, 9, 10
TEKS 1.3B, 1.12A, 1.13

What numbers are missing?

1.

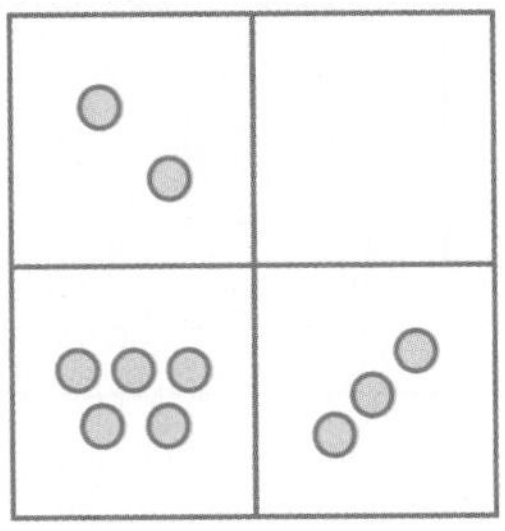

2		2
	3	8
7		

2.

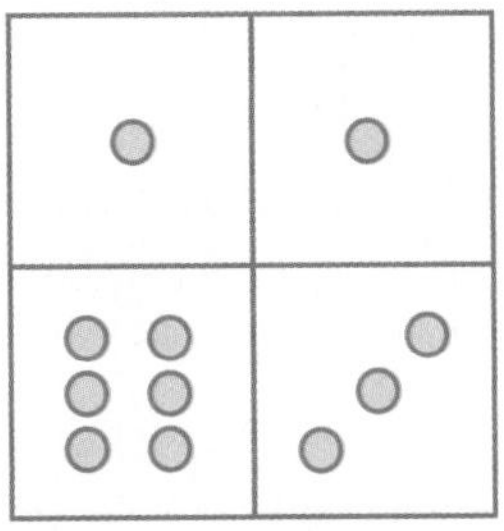

		2
	3	9
7		

3.

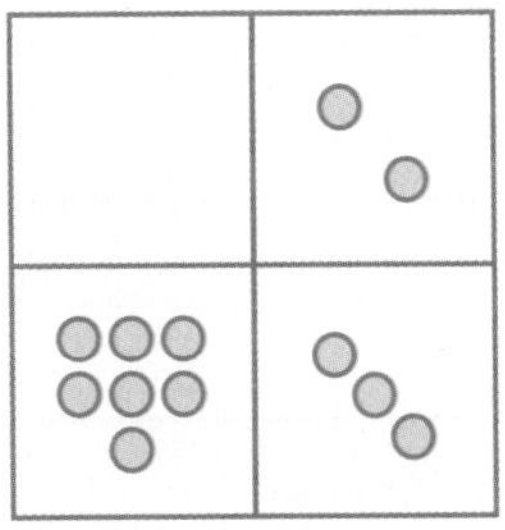

		2
	3	10
7		

NOTE: Completing the Cross Number Puzzles on these pages helps your child develop problem solving skills.

What numbers are missing?

4.

2	1	3
3		6

5.

		4
1		
2	5	

6.

	3	5
	4	10

7.

0		2
3		11

Challenge

8. Jill made her own Cross Number Puzzle. How can you complete it?

		3
	4	
5		

_____

_____

_____

_____

_____

Sums of 10

NCTM Standards 1, 2, 6, 8, 9, 10

TEKS 1.3A, 1.3B

I. Which are sums of 10?

Circle them as fast as you can.

Sums of 10 Search

(3 + 7)	3 + 8	2 + 8
4 + 7	10 + 0	10 + 1
4 + 6	9 + 0	1 + 9
5 + 6	8 + 2	8 + 3
5 + 5	7 + 3	9 + 1
6 + 4	7 + 4	8 + 4
6 + 3	3 + 6	0 + 10

NOTE: Your child is learning to quickly recognize pairs of numbers with a sum of 10. On page 70, children use Cross Number Puzzles to see that numbers can be added in any order and still have the same sum.

© Education Development Center, Inc.

Complete each puzzle. Use the puzzle to complete the number sentence.

2.

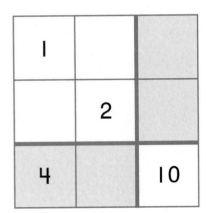

0	1	1
0	9	
0	10	10

__0__ + __10__ = __1__ + _____

3.

	2	2
	8	10

_____ + __8__ = __2__ + _____

4.

1		
	2	
4		10

__4__ + _____ = _____ + _____

5.

		3
	5	
4		10

__4__ + _____ = __3__ + _____

Challenge

6. Make a puzzle for this picture.

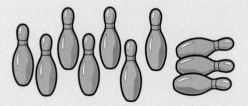

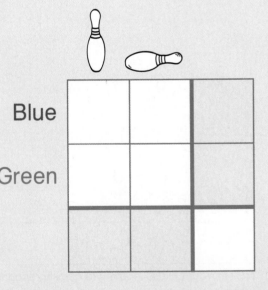

Blue

Green

Name _____ Date _____

Addition Stories

NCTM Standards 1, 2, 6, 7, 8, 9, 10

TEKS 1.3A, 1.11A, 1.13

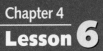

Solve each problem.

1. 5 children played tag.
2 more joined the game.
How many children are
playing tag now?

_____ children

Use words,
numbers, or pictures
to show how you
solved it.

2. Some children are in the classroom.
6 are boys.
3 are girls.
How many children are
in the classroom?

_____ children

NOTE: Your child is learning to solve addition story problems.
You may wish to pose an addition problem about objects or
people in your house and ask your child to solve it.

Write an addition story about the kittens.

3. _____

_____

_____

_____

_____

_____

4. Solve your story problem. Use words, numbers, or pictures to show how you solved it.

Subtraction Stories

NCTM Standards 1, 6, 7, 8, 9, 10

 TEKS 1.3B, 1.13

Use words, numbers, or pictures to show how you solved it.

Solve each problem.

1. 7 children played tag.
   2 went home.
   How many children are playing now?

   _____ children

2. 9 children are in the classroom.
   6 are boys.
   How many are girls?

   _____ girls

NOTE: Your child is learning to solve subtraction story problems.
You may wish to ask your child to compare two groups of objects in your
house. For example, how many more apples do we have than bananas?

✏ **Write a subtraction story about the puppies.**

3. _____

_____

_____

_____

_____

_____

4. Solve your story problem. Use words, numbers, or pictures to show how you solved it.

Name _____ Date _____

Problem Solving Strategy
Guess and Check

NCTM Standards 1, 2, 6, 7, 8, 9, 10
TEKS 1.11A, 1.11B, 1.11C

Understand
Plan
Solve
Check

1. There are 6 birds in a tree.
Some are robins and the rest are sparrows.
There are 2 more sparrows than robins.
How many robins are in the tree?

_____ robins

2. Paul has 7 coins in his pocket.
He only has nickels and pennies.
The coins are worth 15¢.
How many of each type of coin does Paul have?

_____ nickels

_____ pennies

3. Reyna sorts 8 triangles by size.
She has the same number of large
triangles as small triangles.
How many of each size does Reyna have?

_____ small triangles

_____ large triangles

 NOTE: Your child is exploring different ways to solve
problems. Sometimes guessing and checking is an
efficient way to solve a problem.

Problem Solving Test Prep

1. Justin made a pattern with figures.

△ ◻ ◻ ○ △ ◻ ◻ ○ △ ◻ ___ ○

Which figure is missing?

(A) △

(B) ◻

(C) ○

(D) ▭

2. Deandre had 1 nickel and 3 pennies.
Then he earned 10¢.
How much money does he have now?

(A) 10¢

(B) 15¢

(C) 18¢

(D) 23¢

✏ Show What You Know

3. Fiona has some nickels and pennies. She has 11¢ in all. How many of each type of coin could she have?

_____ nickels

_____ pennies

Explain how you know your answer makes sense.

_____

_____

_____

4. Kelly, Josh, and May are in line.
Kelly is in front of Josh.
May is right behind Kelly.
Who is first?

_____ is first.

Explain how you know your answer is correct.

_____

_____

_____

What numbers are missing? Lessons 4 and 5

4.

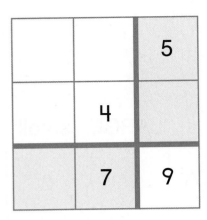

		5
	4	
	7	9

5.

3		8
5		10

Solve each problem. Lessons 6 and 7

6. Marcy picked 4 tulips.
Her mother gave her 5 roses.
How many flowers does Marcy
have now?

_____ flowers

7. There are 10 boys in the park.
Four are on the playground.
The rest are on the grass.
How many are on the grass?

_____ boys

Problem Solving Lesson 8

8. Mark sees some frogs and turtles.
He sees 5 animals in all.
There is 1 more frog than there are turtles.
How many of each kind of animal
does Mark see?

_____ frogs

_____ turtles

Review/Assessment

NCTM Standards 1, 2, 5, 6, 7, 8, 9, 10

I. What figures are missing? Lesson 1
What numbers are missing?

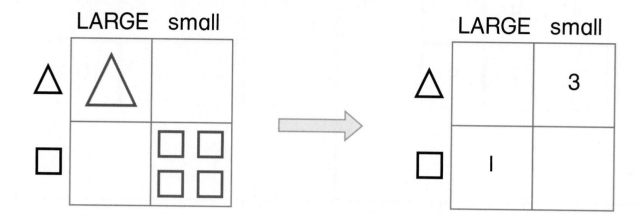

---

2. What numbers are missing? Lesson 2

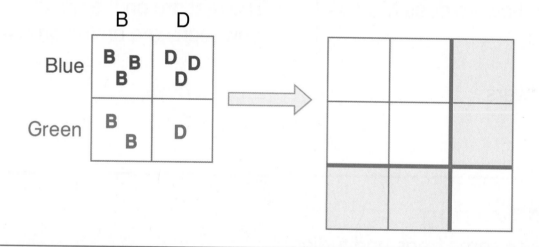

---

3. What numbers are missing? Lesson 3

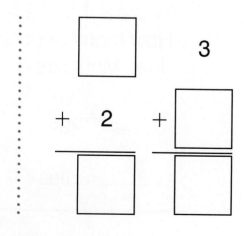

Chapter 5

Working with Tens
Counting by Grouping ✏️

Take two handfuls of counters out of the bag.

STEP 1 Counting

How many counters did you take? _____

How did you find the total?

_____

_____

STEP 2 Counting Another Way

Explain another way you could find the total number of counters.

_____

Use that way to find the total. Compare the two ways of counting.

_____

STEP 3 Counting with Groups of Ten

Make stacks of ten counters. Count the total number again.

Did groups of ten make counting easier? Explain.

_____

_____

_____

_____

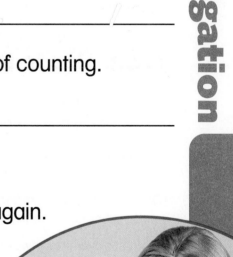

Investigation

> **Dear Family,**
>
> Today we started Chapter 5 of *Think Math!* In this chapter, I will learn that numbers to 50 can be represented as groups of ten and some more. I will also learn to find the value of collections of dimes and pennies, to tell time to the half-hour, and to begin to add with multiples of ten. There are NOTES on some of my pages to explain what I am learning every day.
>
> Here are some activities for us to do together at home. These activities will help me understand numbers to 50.
>
> **Love,**
>
> _____

Family Fun

From Digits to Numbers

Work with your child to make numbers to 50 and locate the numbers on a number line.

- Using index cards, make two sets of number cards 1 to 4. Shuffle the cards and place them face down.

- Draw a number line from 0 to 50 on a piece of paper. Label the line by fives—0, 5, 10, 15, . . . 50.

- The first player draws two cards and arranges them to make a two-digit number. The player then locates and labels that number on the number line. The next player reverses the digits to make a new two-digit number and writes that number on the number line. If the number has already been written on the number line, play moves on to the next round.

- Play continues until no new numbers can be added to the number line.

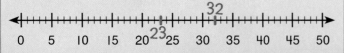

Count the Coins

Work with your child to find the value of a collection of dimes and pennies.

- Put 10 pennies and 10 dimes in a bag. Have your child take some of the coins from the bag without looking and put them in a small pile. Place the remaining coins in a second pile.

- Players look at the dimes and pennies in each pile and guess which collection is worth more. They can indicate their choice by putting a game piece in front of the collection they choose.

- Players work together to find the value of each collection. Players that guessed correctly earn a point.

- Players repeat the game. The first player to earn 5 points wins.

Ten and Some More

NCTM Standards 1, 2, 6, 8, 9, 10
🔺 **TEKS 1.1A, 1.5C**

What numbers are missing?

1.

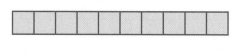

10
+ 1
11

2.

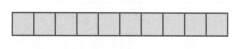

10
+ 5

3.

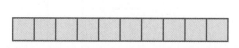

10
+ 8

4.

+ *0*

NOTE: The numbers 11 to 19 can sometimes be challenging for young children. In this lesson your child is learning to say the numbers and identify them as a group of ten and some more.

5. What digits are missing?

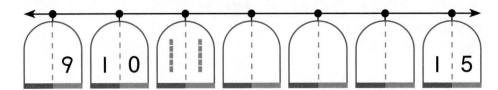

| 9 | 1 0 | | | | | 1 5 |

What numbers are missing?

6. $\boxed{10} + \boxed{6} = \boxed{}$

7. $\boxed{} + \boxed{3} = \boxed{13}$

8. $\boxed{10} + \boxed{} = \boxed{10}$

9. $\boxed{10} + \boxed{2} = \boxed{}$

10.
```
   10
+
─────
   11
```

11.
```
+   7
─────
   17
```

12.
```
   10
+
─────
   14
```

13.
```
+   8
─────
   18
```

Challenge

14. What digits are missing?

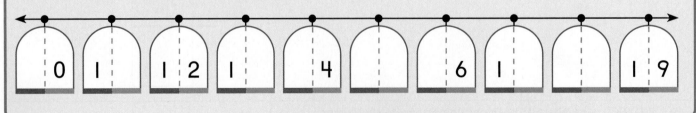

| 0 1 | 1 | 1 2 | 1 | | 4 | | 6 1 | | | 1 9 |

Name _____ Date _____

Lots of Tens and Some More

NCTM Standards 1, 6, 8, 9, 10

🠆 TEKS 1.5C, 1.5D

What is the sum?

1.

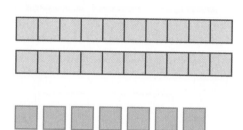

20
+ 7
27

2.

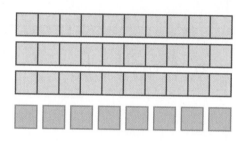

30
+ 8

3.

30
+ 5

4.

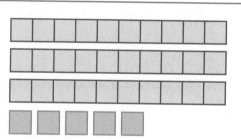

20
+ 9

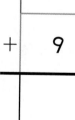

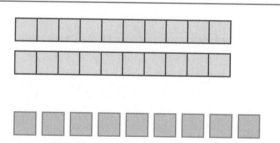

NOTE: Your child is learning to think of the numbers 20 to 50 as a group of tens and some more. Ask your child to use small objects such as pennies or paper clips to show and name these numbers.

80 + 3   **83**   **LXXXIII**   eighty-three **83**

What is missing?

5.

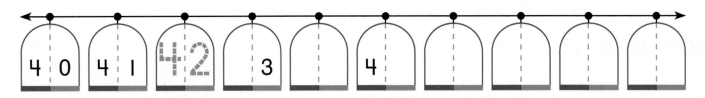

| 4 0 | 4 1 | 42 | 3 | 4 | | | | |

6. 40 + ☐ = 47

7. ☐ + 5 = 45

8. ☐ + 6 = 46

9. 40 + ☐ = 48

10. Draw some sticks of 10 and some loose cubes.
Write an addition sentence to match.

☐ + ☐ = ☐

Challenge
11. What digits are missing?

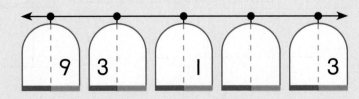

9 3 | | 1 | | 3

Name _____ Date _____

Using Dimes and Pennies

NCTM Standards 1, 4, 6, 8, 9, 10

TEKS 1.1B, 1.1C, 1.12A

How many coins are there? What is the value?

1.

___**3**___ coins

___**30**___ ¢

2.

_____ coins

_____ ¢

3.

_____ coins

_____ ¢

4.

_____ coins

_____ ¢

 5. Look back at Problems 1 to 4.
How are they the same?
How are they different?

_____

_____

NOTE: Your child is learning to count groups of dimes and pennies. Ask your child to use dimes and pennies to show amounts of money up to 99¢.

(D) is for dime, worth 10¢. (P) is for penny, worth 1¢. Write the value in the table.

6.

7.

	How Much?
6.	20 ¢
7.	_____ ¢
8.	_____ ¢
9.	_____ ¢
10.	_____ ¢
11.	_____ ¢

8.

9.

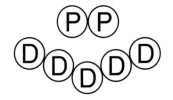

10.

11.

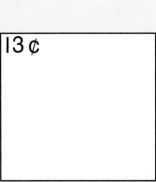

Show 13¢ in 2 different ways.

Problem Solving

12.

13 ¢	13 ¢

Name _____ Date _____

Tens and Time

NCTM Standards 1, 4, 6, 10

TEKS 1.8B

What time is it?

1.

2.

3.

4.

NOTE: Your child is learning to tell time to the half-hour. For times on the half-hour, have your child describe the positions of the hour hand and minute hand on a clock and tell what time it is.

Draw the hour hand and the minute hand.

5.

2 hours later →

The time is 8:30.
The time is 10:30.

6.

3 hours later →

The time is 1:30.
The time is 4:30.

7.

I hour later →

The time is 6:30.
The time is ___:___.

Problem Solving

8. A movie begins at 11:00.
It is two and one-half hours long.
What time does it end?

The movie ends at ___:___.

88

44 + 44

© Education Development Center, Inc.

Chapter 5
Lesson 5

Tens on the Number Line

NCTM Standards 1, 2, 6, 9, 10

🔻 **TEKS 1.5D**

Write the numbers to match each jump.

1.

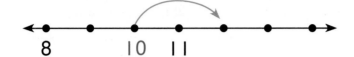

8   10 11

2.

8   11

3.

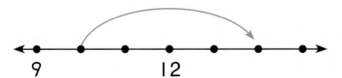

9   12

4.

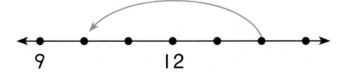

9   12

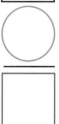

NOTE: Your child is looking at addition and subtraction as forward or backward jumps on a number line. Ask your child to explain the meaning of each jump.

Draw the jump. Complete the number sentence.

5.

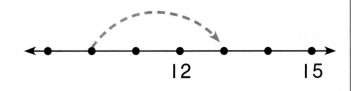

$$\boxed{10} + \boxed{3} = \boxed{13}$$

6.

19   21   23

$$\boxed{20} + \boxed{2} = \boxed{}$$

7.

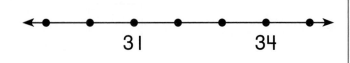

31   34

$$\boxed{} + \boxed{5} = \boxed{35}$$

8.

41   46

$$\boxed{} + \boxed{3} = \boxed{43}$$

9.

28   33

$$\boxed{30} + \boxed{} = \boxed{31}$$

10.

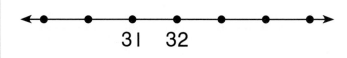

31 32

$$\boxed{30} + \boxed{} = \boxed{34}$$

Challenge

11. Write a number sentence to match the jump.

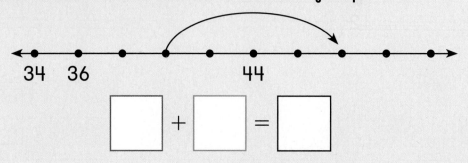

34 36   44

$$\boxed{} + \boxed{} = \boxed{}$$

Using the Number Line to Solve Problems

NCTM Standards 1, 2, 6, 7, 9, 10

🔹 **TEKS 1.5D, 1.12A**

Draw the jump. Complete the number sentence.

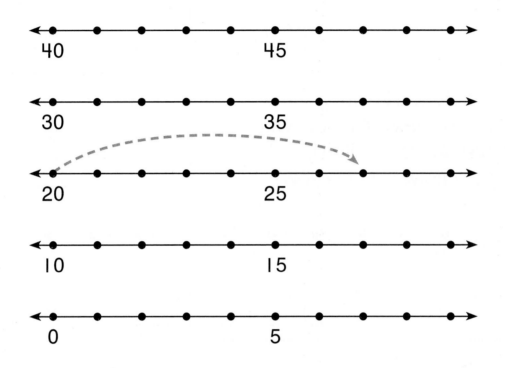

1. | 20 | + | 7 | = | 27 |

2. | 30 | + | | = | 34 |

3. | 10 | + | | = | 16 |

4. | | + | 1 | = | 41 |

5. | | + | 5 | = | 45 |

6. | 20 | + | | = | 23 |

7. | 30 | + | | = | 38 |

8. | | + | 2 | = | 12 |

Draw the jump to solve each problem.
Complete the number sentence.

9. There are 20 children in class.
3 more children come late.
How many children are in class now?   _____ children

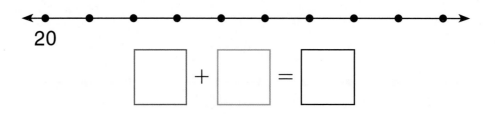

20

☐ + ☐ = ☐

---

10. There are 30 red flowers.
There are some yellow flowers too.
There are 38 flowers in all.
How many flowers are yellow?   _____ flowers

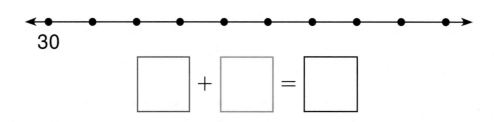

30

☐ + ☐ = ☐

Challenge

11.
20
+ 0
_____

12.
40
+ 10
_____

13.
20
+ 20
_____

14.
10
+ 20
_____

15. What patterns do you see?

_____

_____

Name _____ Date _____

Modeling Numbers in Different Ways

NCTM Standards 1, 2, 4, 6, 7, 9, 10
TEKS 1.1B, 1.1C, 1.12A

1. Which show 32?

$30 + 2$

$23 + 10$

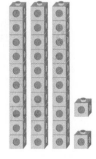

$42 + 10$

$10 + 10 + 3$

$10 + 10 + 10 + 2$

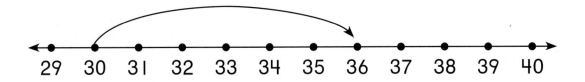

29 30 31 32 33 34 35 36 37 38 39 40

---

2. Pick one model from Problem 1 that does not show 32. How can you change it to show 32? Draw or write below.

NOTE: Your child has been learning different ways to represent a number. Ask your child to name a number between 10 and 50 and then show that number in as many different ways as he or she can.

What number does each show?
Show the number another way.

3.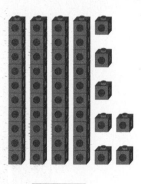

4.

5. 10 + 10 + 10 + 8

Problem Solving

6. I am a two-digit number.
I have fewer tens than ones.
The number of ones is less than 5.
I am less than 40.

What number could I be? _____

Name _____   Date _____

Problem Solving Strategy
Draw a Picture

NCTM Standards 1, 4, 6, 7, 8, 9, 10

🤚 TEKS 1.4, 1.11A, 1.11B, 1.11C, 1.13

Understand

Plan

Solve

Check

1. I have 2 dimes and 3 pennies. My brother gives me 1 dime and 2 pennies. How much money do I have now?

_____ ¢

2. Ben is next to Lisa. Lisa is next to Jerry. Who is in the middle?

_____

3. Anna makes a necklace with beads. She uses 1 red bead then 2 blue beads. She follows the same pattern. What color is the 7th bead?

_____

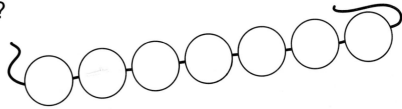

© Education Development Center, Inc.

NOTE: Your child is learning to use the problem solving strategy, *draw a picture*, to solve problems.

Problem Solving Test Prep

1. Jo found the sum of some tens and 4 ones.

   Which could be the sum?

   (A) 4   (C) 42

   (B) 24   (D) 46

2. A lobster has 10 legs. How many legs are on 3 lobsters?

   (A) 3 legs   (C) 30 legs

   (B) 13 legs   (D) 33 legs

 Show What You Know

3. Ana eats lunch at 11:30. Draw the hour hand and the minute hand to show this time.

   Describe where the hands belong.

   _____

   _____

   _____

   _____

4. I have some dimes and pennies. I have more pennies than dimes. I have fewer than 9 pennies.

   What is the most money I could have?

   _____¢

   Explain how you know your answer is correct.

   _____

   _____

   _____

   _____

© Education Development Center, Inc.

Chapter 5 # Review/Assessment
NCTM Standards 1, 2, 6, 7, 8, 9, 10

What numbers are missing? Lesson 1, 2, and 7

1.

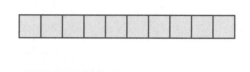

10
+ 4

2.

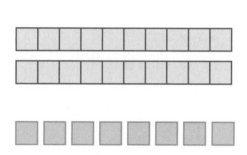

20
+ 8

3.

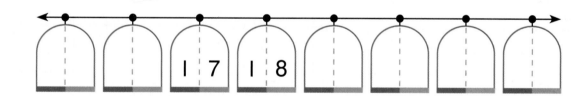

| 1 | 7 | 1 | 8 |

How many coins are there? What is the value? Lesson 3

4.

_____ coins

_____ ¢

5.

_____ coins

_____ ¢

What time is it? Lesson 4

6.

_____:_____

7.

_____:_____

Draw the jump. Complete the number sentence. Lesson 5, 6, and 7

8.

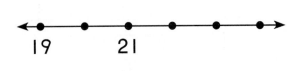

| 20 | + | 3 | = | [ ] |

9.

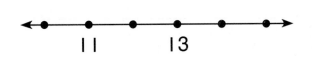

| 10 | + | [ ] | = | 15 |

10.

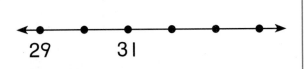

| [ ] | + | 4 | = | 34 |

11.

| 20 | + | [ ] | = | 21 |

Problem Solving Lesson 8

12. Julio has 3 dimes and 2 pennies.
He finds 1 more dime and 4 more pennies.
How much money does he have now?

_____¢

Name _____

Data and Probability
Flipping a Two-Color Counter

You need
- two-color counter
- red crayon
- yellow crayon

Flip the counter to see which color
it will land on.

STEP 1 **Flipping the Counter**

Did you flip red or yellow? _____

STEP 2 **Recording Results**

Color 1 box to show the color you flipped.

What do you think will happen if you flip the counter 10 times?

_____

_____

STEP 3 **Gathering Data**

Flip the counter 9 more times.
Color a box to show each flip.

How many times did you flip red? _____

How many times did you flip yellow? _____

What do you think you would flip next?

_____

_____

Investigation

Dear Family,

Today we started Chapter 6 in *Think Math!* In this chapter, I will collect, organize, and graph data and describe how likely things are to happen. There are NOTES on the Lesson Activity Book pages to explain what I am learning every day.

Here are some activities for us to do together at home. These activities will help me understand data and probability.

Love,

_____

Family Fun

Tossing Sums

Work with your child to practice sums to 12.

- Gather two number cubes labeled 1 to 6 and a sheet of paper.

- Have your child toss the number cubes and find the sum of the numbers tossed.

- Check the sum and record it on a sheet of paper.

- Repeat this for at least 10 tosses, making a list of each sum tossed.

- Ask your child "Which number came up most often? Which came up least often?"

Sums Tossed		
8	5	3
5	2	9
7	8	8
4		

Coins in a Bag

Work with your child to record data in a table.

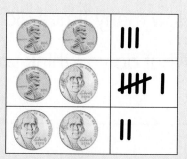

- Gather 2 pennies, 2 nickels, and a paper bag. Place all 4 coins in the bag.

- Ask your child what combinations of two coins could be pulled out of the bag. Work together to make a table to show all of the possible combinations.

- Have your child pull 2 coins out of the bag without looking. Use tally marks to record the combinations of coins he or she pulled. Return the coins to the bag and repeat.

penny	penny	III
penny	nickel	HHI I
nickel	nickel	II

- After a few rounds, ask your child which combinations were pulled most often.

Collecting and Tallying Data

NCTM Standards 1, 5, 6, 7, 8, 9, 10
TEKS 1.10A

Joel asked his friends if they have a dog.

Do you have a dog?		
yes	no	no
no	yes	yes
no	yes	no
yes	no	yes
no	no	no

‖‖‖ means 5.

1. Use tally marks to show the data.

Yes	No
‖‖‖ l	

2. How many friends have dogs?

_____ friends

3. How many friends do not have dogs?

_____ friends

 4. Do most of his friends have dogs?
Explain how you know.

_____

_____

_____

NOTE: Your child is learning to use tally marks to record
and analyze data. Ask your child to use tally marks to keep
track of the birds or cars they see outside.

Lisa asked her friends if they like to swim.

Do you like to swim?	
Yes	**No**
卌 卌 III	卌 III

5. How many friends like to swim?

13 _____ friends

6. How many friends do not like to swim?

_____ friends

7. How many friends did Lisa ask?

_____ friends

8. How many more friends like to swim than do not like to swim?

_____ more friends

 9. Explain how you found the answer to Question 8.

_____

_____

_____

Problem Solving

10. Max asked 10 friends if they like pizza. 6 more children like pizza than do not like pizza.

Draw tally marks to show the results.

Do you like pizza?	
Yes	**No**

Making Graphs with Objects and Pictures

NCTM Standards 1, 5, 6, 7, 8, 9, 10

TEKS 1.10A, 1.13

Kylie sorts her buttons by shape.

1. There are ____**4**____ circle buttons.

2. There are more _____ buttons than circle buttons.

3. There are fewer _____ buttons than star buttons.

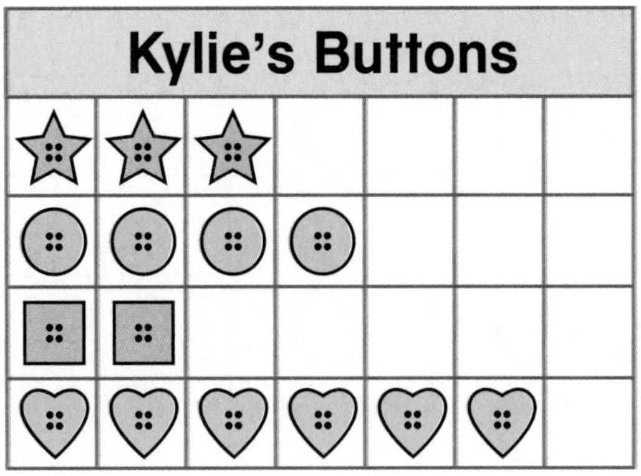

Kylie's Buttons					
⭐	⭐	⭐			
⬤	⬤	⬤	⬤		
▪	▪				
♥	♥	♥	♥	♥	♥

4. There are _____ buttons in the graph.

Jim sorts blocks by size and shape.

5. There are _____ small triangles.

6. There are more _____

   than _____ .

7. There are fewer _____ than any other kind of block.

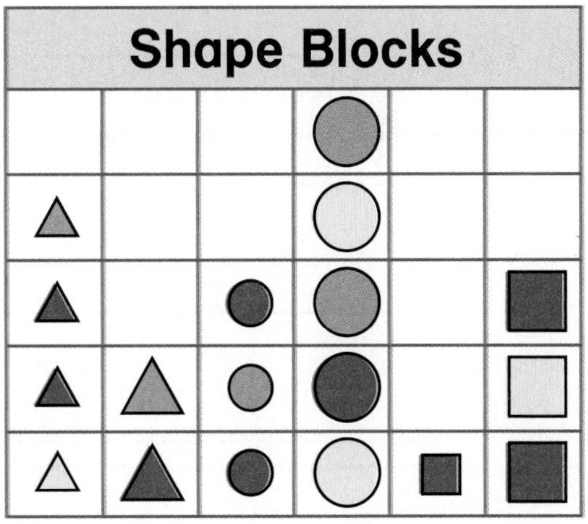

Shape Blocks					
			⬤		
△			◯		
△		●	⬤		■
△	△	●	⬤		□
△	▲	●	◯	▪	■

8. There are _____ blocks in the graph.

NOTE: Your child is learning to sort objects and to represent data in graphs. Ask your child how many small triangles there are in the picture graph above.

9. Use the clues to color the graph.

Bouncy Balls			
◯			
◯			
◯			◯
◯	◯		◯
◯	◯		◯
◯	◯	◯	◯
◯	◯	◯	◯
◯	◯	◯	◯

Clues
There are 8 green balls.
There are 3 red balls.
There are more blue balls than yellow.

Each column in the graph shows one color.

10. How did you know which circles to color blue and which to color yellow?

_____

_____

_____

_____

Challenge

11. Use the graph above to complete the sentence.

There are 2 more _____ balls

than _____ balls.

Chapter 6
Lesson 3 # Making Graphs with Pictures and Symbols

NCTM Standards 1, 2, 5, 6, 7, 8, 9, 10
🐟 TEKS 1.9B, 1.10A, 1.13

Some children chose their favorite fish. The table shows their choices.

Which of these fish do you like best?	
Goldfish	⊮⊮⊮ I
Guppy	II
Angelfish	IIII
Neon	I

1. Use the data in the table to make a graph.

Fish We Like

☺			
Goldfish	Guppy	Angelfish	Neon

Key: Each ☺ stands for 1 child's choice.

Draw ☺ to show each child's choice.

2. Which fish did the most children choose?

_____

3. How is the graph like the table? How is it different?

_____

_____

_____

 NOTE: Your child is learning to make and use pictographs. Ask your child to tell how many children chose angelfish in the graph above.

© Education Development Center, Inc.

Matthew asked his classmates to name their favorite colors. The graph shows his data.

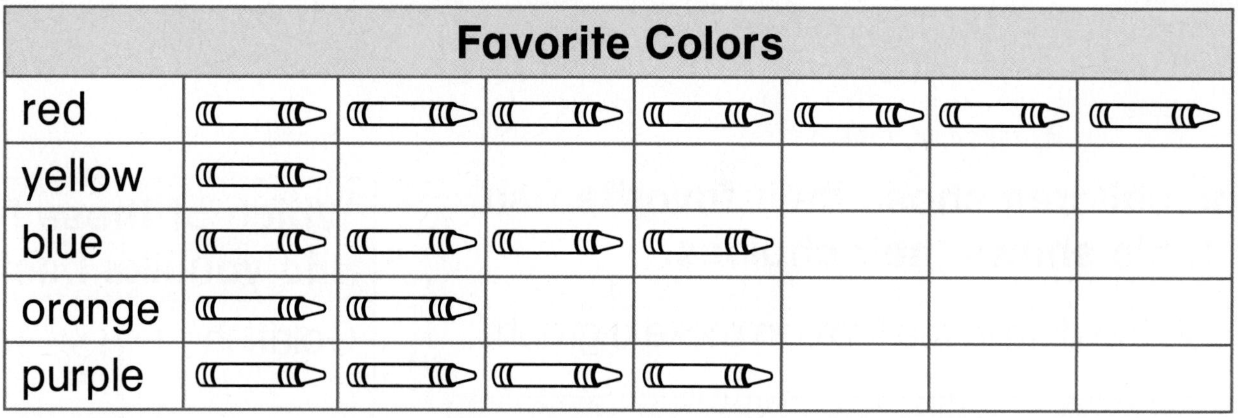

Favorite Colors							
red	✏	✏	✏	✏	✏	✏	✏
yellow	✏						
blue	✏	✏	✏	✏			
orange	✏	✏					
purple	✏	✏	✏	✏			

Key: Each ✏ stands for 1 child's choice.

4. How many children chose purple? ___4___ children

5. Which color did the fewest children choose? _____

6. How many more children chose red than blue?
Explain how you know.

_____

_____

_____

Problem Solving

7. Use the clues to complete the graph.

- 3 more children chose baseball than football.
- 2 more children chose soccer than football.

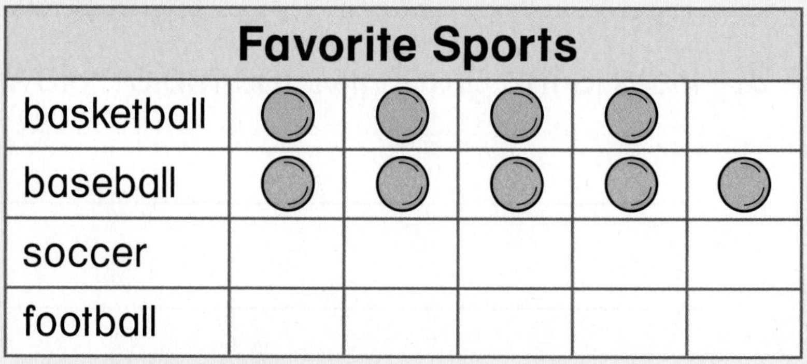

Favorite Sports					
basketball	⬤	⬤	⬤	⬤	
baseball	⬤	⬤	⬤	⬤	⬤
soccer					
football					

Key: Each ⬤ stands for 1 child's choice.

Bar Graphs and Probability

NCTM Standards 1, 5, 6, 7, 8, 9, 10
TEKS 1.9B, 1.10A, 1.13

**Trevor tossed a number cube 20 times.
The tally table shows his results.**

Number Tossed	Number of Tosses
1	II
2	IIII
3	III
4	HHT
5	III
6	III

1. Use the tally table to complete the bar graph.

Number Cube Toss

Number of Tosses

5
4
3
2
1
0

1   2   3   4   5   6
Number Tossed

---

2. How did you know how high to color each bar in the graph?

_____

_____

_____

3. Which number did Trevor toss the most? _____

4. Which number did Trevor toss the least? _____

NOTE: Your child is learning to make and analyze bar graphs. Ask your child to use the bar graph above to tell how many times Trevor tossed the number 6.

5. What sums can you get when you toss two number cubes? Write the missing sums.

+	1	2	3	4	5	6
1	2					
2				6		
3						
4			7			
5						
6						

6. What are all the ways to toss a sum of 4?

Challenge

7. What are all the ways to toss a sum of 7?

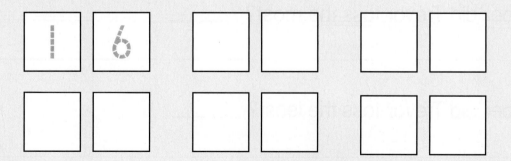

Chapter 6
Lesson 5

Investigating Probability

NCTM Standards 1, 5, 6, 7, 8, 9, 10
TEKS 1.9A, 1.9B, 1.10A, 1.10B

I. Toss two number cubes.
Color a box for the sum.
Which numbers win? Have fun!

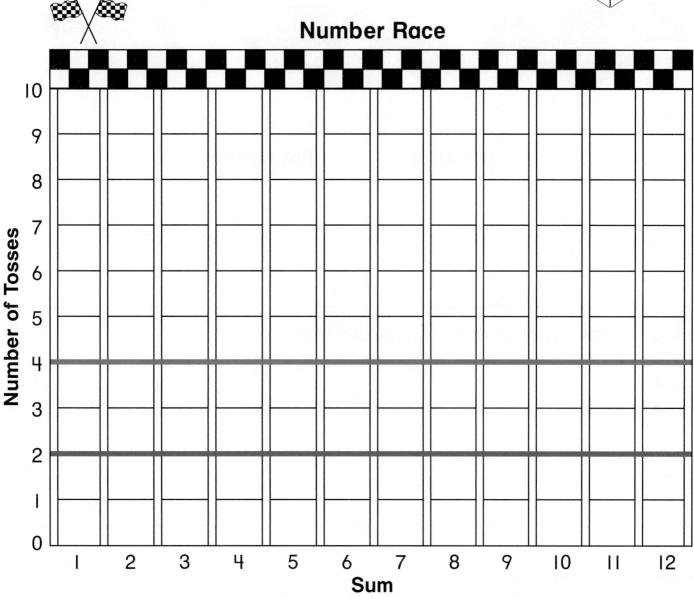

Number Race

Number of Tosses (y-axis: 0, 1, 2, 3, 4, 5, 6, 7, 8, 9, 10)

Sum (x-axis: 1, 2, 3, 4, 5, 6, 7, 8, 9, 10, 11, 12)

Winner! **Winner!** **Winner!**

NOTE: Your child is learning to describe events as *certain,*
uncertain, likely, and *unlikely.* Ask your child to name events
that are either *likely* or *unlikely.*

Is it *possible* or *impossible?*

2. A car will drive by the school.

3. A cow will fly over your house.

4. A bird will land in front of a house.

possible

impossible

---

5. Draw a picture of an impossible event.

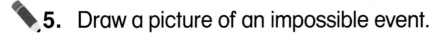

Challenge

Is it *certain, likely,* or *unlikely?* Draw lines to match.

6. You will wear matching socks tomorrow.   certain

7. The sun will rise tomorrow morning.   likely

8. All of your classmates will go to the dentist today.   unlikely

Name _____ Date _____

Problem Solving Strategy
Make a Table ✏

NCTM Standards 1, 2, 5, 6, 7, 8, 9, 10

🔊 TEKS 1.11B, 1.11C

Understand
Plan
Solve
Check

Complete the table to solve the problem.

1. How many more children chose apples than grapes?

Favorite Kind of Fruit	
apples	grapes
bananas	bananas
apples	apples
grapes	apples
bananas	grapes
bananas	apples

Favorite Kind of Fruit	
🍎 apples	
🍇 grapes	
🍌 bananas	

_____ more children

2. What are all the ways to add two numbers to get a sum of 10?

$$\square + \triangle = 10$$

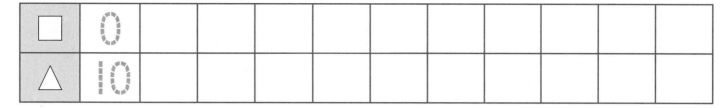

$\square$	0										
$\triangle$	10										

_____ ways

NOTE: Your child is exploring different ways to solve problems. Making a table can help to organize the information you need to solve a problem.

© Education Development Center, Inc.

Problem Solving Test Prep

1. Kelly buys 8 pairs of socks.
Some are white.
The rest are black.
She has 2 more white pairs
than black.
How many pairs are white?

(A) 3   (C) 8

(B) 5   (D) 10

2. Paolo has 18¢.
Erasers cost 5¢ each.
How many erasers can
he buy?

(A) 3   (C) 18

(B) 4   (D) 20

 Show What You Know

3. Derek makes a pattern
with squares.

□ □□ □□□ □□□□ □□□□□

How many squares
are in the next figure? _____
Explain.

_____

_____

_____

_____

_____

_____

4. Ann has some dimes
and pennies. She has 32¢.
How many dimes and pennies
could she have?

_____ dimes _____ pennies

Explain.

_____

_____

_____

_____

_____

_____

Chapter 6 Review/Assessment
NCTM Standards 1, 2, 5, 6, 7, 8, 9, 10

Claire asked her classmates if they like apples. Lesson 1

1. Use tally marks to show the data.

Do you like apples?		
yes	no	yes
yes	yes	yes
no	yes	no
yes	no	yes
no	yes	yes

Yes	No

2. How many classmates like apples? _____ classmates

Use the graph for Problems 3 and 4. Lesson 2

3. There are _____ circles.

4. There are more _____ than squares.

Shapes

Use the graph for Problems 5 and 6. Lesson 3

5. Which pet did most children choose?

_____

6. How many more children chose dogs than cats?

_____ more children

Favorite Pets

cats	☺ ☺ ☺ ☺ ☺		
dogs	☺ ☺ ☺ ☺ ☺ ☺ ☺		
hamsters	☺ ☺ ☺		

Key: Each ☺ stands for 1 child's choice.

David pulled cubes out of a bag without looking.
The tally table shows his results. Lesson 4

7. Use the table to complete the bar graph.

Cubes Pulled from a Bag

Colors (vertical axis)

green	▓	
yellow		
orange		

0 1 2 3 4 5 6 7 8
Number of Times Pulled

Colors	Number of Times Pulled
green	IIII
yellow	IIII III
orange	II

8. Which color did David pull the fewest times? _____

9. How many times did David pull out a yellow cube? _____ times

Is it *certain, likely,* or *unlikely*?
Draw lines to match. Lesson 5

10. All of your classmates will wear
the same color shirt tomorrow.   certain

11. It will rain sometime this year.   likely

12. Tomorrow it will be today.   unlikely

Problem Solving Lesson 6

13. Mr. Lee has 4 tables in his
classroom. He wants 4 chairs
around each table. How many
chairs does he need?

tables	1	2	3	4
chairs				

_____ chairs

Working with Larger Numbers

How Much is 100?

You need
• pennies

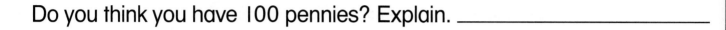

Put a pile of pennies on your desk.

STEP **1** Estimating

Do you think you have 100 pennies? Explain. _____

_____

STEP **2** Finding Ways to Count

Count the pennies in your pile.

Describe what you did as you counted the coins. _____

_____

_____

_____

_____

STEP **3** Comparing Groups

Count out 100 pennies.

Does this group have more or fewer pennies than your pile? Explain.

_____

_____

_____

Investigation

Dear Family,

    Today we started Chapter 7 in *Think Math!* In this chapter, I will learn to recognize, add, and subtract numbers to 100. I will learn number words for larger numbers. I will also learn about the quarter and find the value of a collection of coins. There are NOTES on the Lesson Activity Book pages to explain what I am learning every day.

    Here are some activities for us to do together at home. These activities will help me understand money and strategies for counting.

Love,

_____

Family Fun

Number Name Concentration

Work with your child to practice recognizing number names.

- Use index cards or slips of paper to make number and number name cards. Make 16 matching pairs of numbers and number names for any numbers from 11 to 99.

- Shuffle the cards and lay them face down on the table.

- Take turns flipping over two cards at a time. Try to find a number and its matching number name. If you find a match, put the cards aside in your pile. If you do not find a match, turn the cards face down again.

- Play until all of the cards have been matched. The player with the most cards at the end of the game wins.

Add Ten, Subtract Ten

Work with your child to use mental math to add and subtract in everyday situations.

- Encourage your child to find two-digit numbers around your neighborhood. For example, you might point out two-digit prices on items at a store. Have your child read the number aloud.

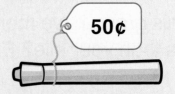

- Have your child tell you what ten more than the number is and what ten less than the number is. Your child might want to draw a picture or use buttons or pennies to help. With practice, your child will be able to add or subtract ten automatically.

Identifying Rules

NCTM Standards 1, 2, 6, 7, 8, 9, 10

TEKS 1.5D, 1.13

Name _____ Date _____

What is the rule?

1.

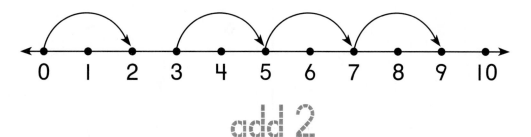

add 2

_____

2.

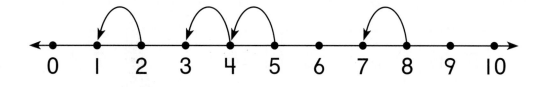

_____

3.

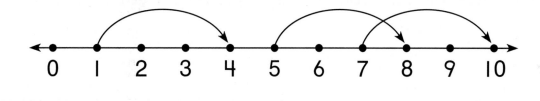

_____

4. Make your own. Draw jumps on the number line.
What is the rule?

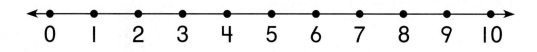

_____

NOTE: Your child is learning to identify patterns in jumps on
a number line. Have your child explain how to find the rule
for Problem 4.

© Education Development Center, Inc.

Draw the missing jumps. Complete each table.

5.

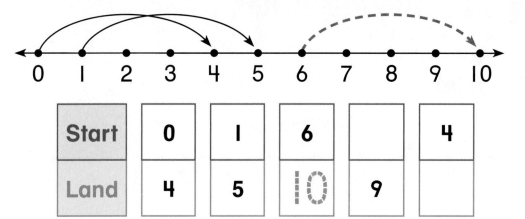

Start	0	1	6		4
Land	4	5	10	9	

6.

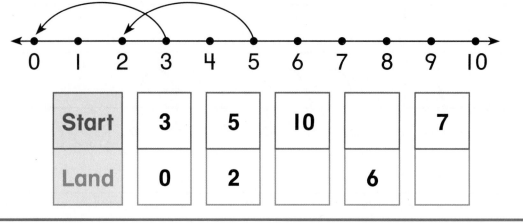

Start	3	5	10		7
Land	0	2		6	

✏ **7.** What is the rule for Problem 6? Explain how you know.

_____

_____

_____

_____

Challenge
8. Find the rule. Complete the table.

Start	2	5	4	0		10		12
Land	7	10	9		6		20	

Identifying Rules with Larger Numbers

NCTM Standards 1, 2, 6, 7, 8, 9, 10
TEKS 1.3B, 1.4

Draw the missing jumps.
Complete each table.

1.

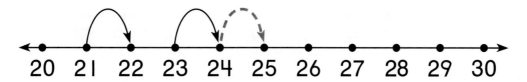

Start	21	23	24	27	
Land	22	24	25		23

2.

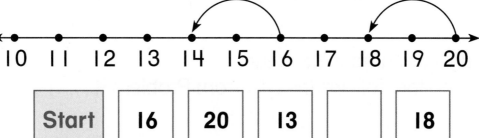

Start	16	20	13		18
Land	14	18		13	

3.

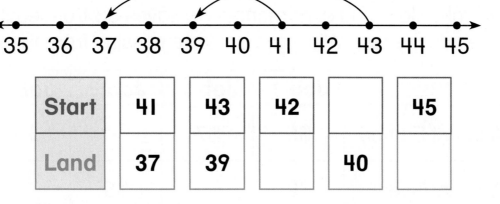

Start	41	43	42		45
Land	37	39		40	

NOTE: Your child is learning to identify patterns in number line jumps. Have your child tell you the rule for each problem on this page.

© Education Development Center, Inc.

4. Complete the table. Use the jumps on A to D.

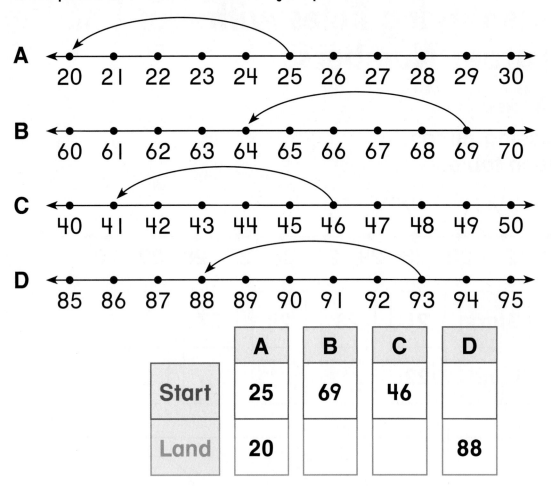

	A	**B**	**C**	**D**
Start	25	69	46	
Land	20			88

---

5. Make your own number line chunk.
Show a jump that follows the rule from Problem 4.

Problem Solving

6. Harry rakes lawns to earn money. He saves $3 from each job. He spends the rest. What numbers are missing from the table?

	Job 1	**Job 2**	**Job 3**	**Job 4**
Amount Earned	$5	$7		$4
Amount Spent	$2		$5	

Adding Ten on the Number Line Hotel

NCTM Standards 1, 2, 6, 7, 8, 9, 10

What is the missing number?

1.

$$3 + \boxed{10} = \boxed{13}$$

2.

$$13 + \boxed{} = \boxed{23}$$

3.

$$23 + \boxed{} = \boxed{33}$$

4.

$$18 + \boxed{} = \boxed{28}$$

5.

$$26 + \boxed{} = \boxed{36}$$

6.

$$31 + \boxed{} = \boxed{41}$$

NOTE: Your child is learning to add ten using the Number Line Hotel. Ask your child to describe the rule for the problems on this page.

What is the missing number?
Draw the missing jump.

7. $\boxed{12}$ + $\boxed{10}$ = $\boxed{22}$

8. $\boxed{27}$ + $\boxed{10}$ = $\boxed{}$

9. $\boxed{0}$ + $\boxed{10}$ = $\boxed{}$

10. $\boxed{37}$ + $\boxed{10}$ = $\boxed{}$

11. $\boxed{}$ + $\boxed{10}$ = $\boxed{43}$

12. $\boxed{}$ + $\boxed{10}$ = $\boxed{16}$

13. $\boxed{}$ + $\boxed{10}$ = $\boxed{31}$

14. $\boxed{}$ + $\boxed{10}$ = $\boxed{19}$

Challenge

What is the missing number?

15. $\boxed{58}$ + $\boxed{10}$ = $\boxed{}$

16. $\boxed{87}$ + $\boxed{10}$ = $\boxed{}$

17. $\boxed{}$ + $\boxed{10}$ = $\boxed{76}$

18. $\boxed{}$ + $\boxed{10}$ = $\boxed{85}$

Name _____ Date _____

Subtracting Ten on the Number Line Hotel

NCTM Standards 1, 2, 6, 7, 8, 9, 10

What is the missing number?

1.

$$13 - \boxed{10} = 3$$

2.

$$23 - \boxed{} = 13$$

3.

$$33 - \boxed{} = 23$$

4.

$$28 - \boxed{} = 18$$

5.

$$36 - \boxed{} = 26$$

6.

$$41 - \boxed{} = 31$$

NOTE: Your child is learning to subtract ten using the Number Line Hotel. Ask your child to describe the rule for the problems on this page.

What is the missing number?
Draw the missing jump.

7. | 22 | − | 10 | = | 12 |

11. | | − | 10 | = | 33 |

8. | 37 | − | 10 | = | |

12. | | − | 10 | = | 6 |

9. | 10 | − | 10 | = | |

13. | | − | 10 | = | 21 |

10. | 40 | − | 10 | = | |

14. | | − | 10 | = | 9 |

Challenge
What is the missing number?

15. | 68 | − | 10 | = | |

17. | | − | 10 | = | 66 |

16. | 97 | − | 10 | = | |

18. | | − | 10 | = | 75 |

Adding and Subtracting with Larger Numbers

NCTM Standards 1, 2, 6, 7, 8, 9, 10

1. What numbers are missing?
Draw the missing jumps.

Start	13	32	3	24	37	45	
Jump Forward	1	3	2	1	2		2
Land	14	35				48	12

NOTE: Your child is investigating functions with two inputs by filling in missing numbers in a table. Have your child explain the rule for the table above.

What numbers are missing?

2.

Start	12	25	36	43		14	
Jump Forward	3	5	2		10	6	10
Land	15	30		47	38		30

3.

Start	12	25	36	43		14	
Jump Back	3	5	2		10	6	10
Land	9	20		43	38		30

4. What is the rule for Problem 3? Explain how you know.

_____

_____

_____

_____

Challenge

5. What numbers are missing?

		2	4	3	5		9
→		3	8	6		1	
Room Number		23	48		57	71	90

Chapter 7
Lesson 6

Modeling Numbers to 99

NCTM Standards 1, 2, 6, 7, 8, 9, 10

TEKS 1.1B

I. Draw a line to match.

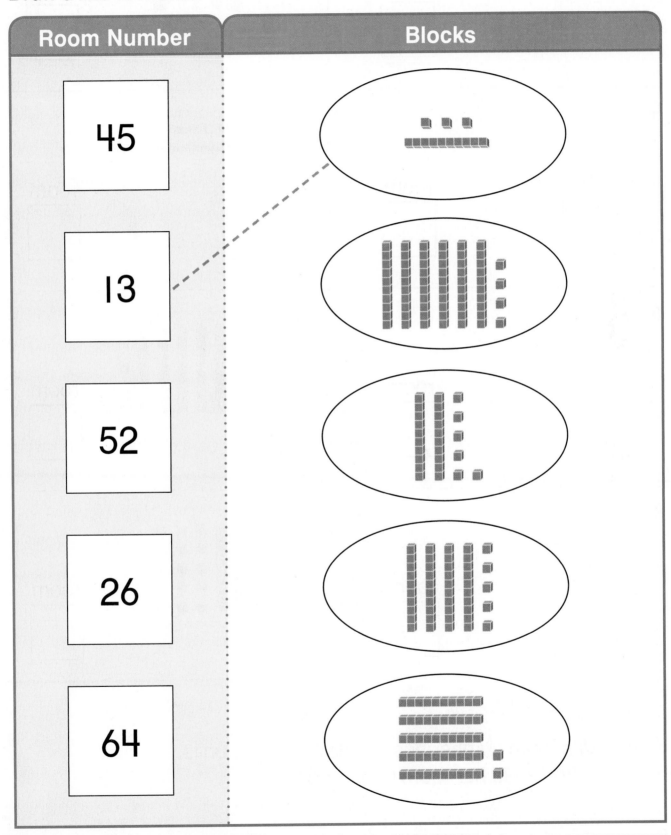

Room Number	Blocks
45	
13	
52	
26	
64	

NOTE: Your child is learning to recognize the base-ten block representations of numbers. Ask your child what the rods and units show.

What is the number?

2.

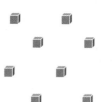

room

3̣1̣

3.

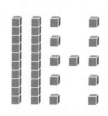

room

4.

room

5.

room

6.

room

7.

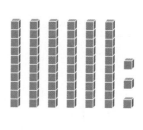

room

8.

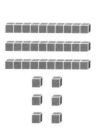

room

9.

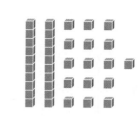

room

Problem Solving

10. Vi's room number has 6 rods and some units.
What could her room number be?

_____

Name _____ Date _____

Numbers to 100 and Beyond

NCTM Standards 1, 2, 6, 7, 8, 9, 10

What numbers are missing?

1. 67 → → ↑ ↑

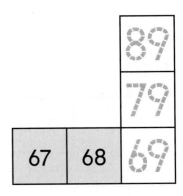

89
79
67 | 68 | 69

Think of the Number Line Hotel!

2. 83 → ↑ → →

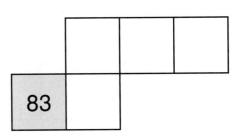

83

3. 94 → ↑ ↑ ← ↑

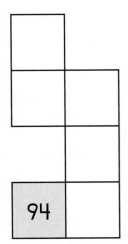

94

4. 71 ↑ → ↑ → ↑

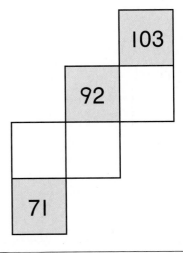

103
92
71

5. 97 → ↑ ↑ ↑ →

128 | 129
118
97

🏠 **NOTE:** Your child is using number patterns to explore numbers beyond 100. You might give your child a starting number and a path of arrows and ask for the landing number.

© Education Development Center, Inc.

What numbers are missing?

6.

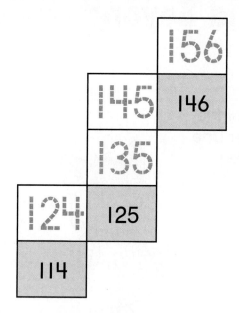

7.

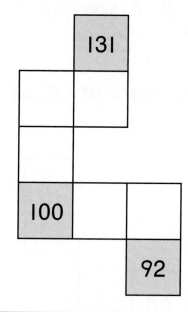

8.

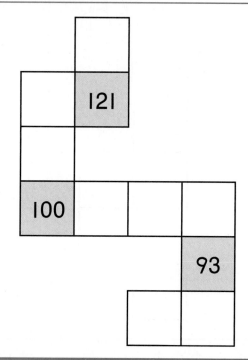

9.

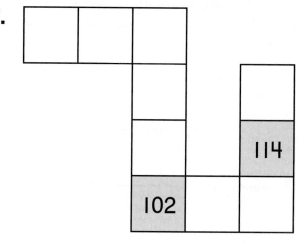

Challenge

10. Write the landing number.

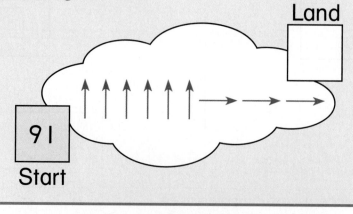

Land

91
Start

Name _____ Date _____

Connecting Numbers and Words

NCTM Standards 1, 2, 6, 7, 8, 9, 10
TEKS 1.1D, 1.13

I. Draw lines to match.

twenty   12

twelve   4

ten   14

four   20

fourteen   10

forty   13

thirty   40

thirteen   30

fifty-seven   75

eighty-nine   17

seventy-five   57

seventeen   99

sixty-one   89

ninety-nine   61

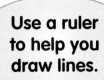

Use a ruler
to help you
draw lines.

NOTE: Your child is learning to match numbers with word
names and vice versa. You may ask your child to explain how a
word name can tell how many tens and ones are in a number.

Write each number.

2. sixteen   `16`

3. fifteen   ☐

4. forty-seven   ☐

5. twenty-nine   ☐

6. thirty-six   ☐

7. sixty-three   ☐

8. fifty-eight   ☐

9. seventy-four   ☐

10. ninety-two   ☐

11. eleven   ☐

Write each word name.

Use the word names on the page to help you.

12. 56 _____

13. 22 _____

14. How many tens and ones are there in forty-three? Explain how you know.

_____

_____

_____

_____

Challenge

15. What number is missing?

Twenty-seven plus thirteen equals _____

Name _____ Date _____

Introducing the Quarter

NCTM Standards 1, 2, 6, 7, 8, 9, 10

🔷 TEKS 1.1C

1. Draw lines to match.

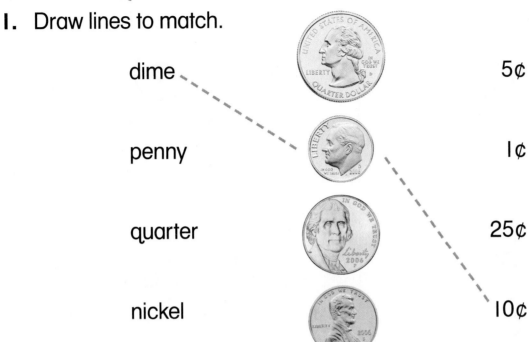

dime   5¢

penny   1¢

quarter   25¢

nickel   10¢

What is the value?

2.

50 ____ ¢

3.

_____ ¢

4.

_____ ¢

5.

_____ ¢

🏠 **NOTE:** Your child is learning to find the value of a collection of coins from 50¢ to 100¢ using quarters, dimes, nickels, and pennies. You may want to give your child some coins and ask what the value is.

What is the value?

6.

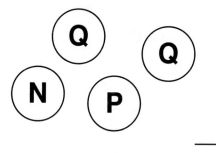

_____ ¢

7.

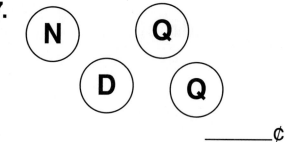

_____ ¢

8.

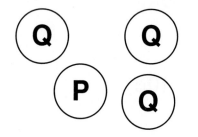

_____ ¢

9.

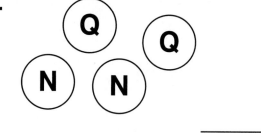

_____ ¢

10.

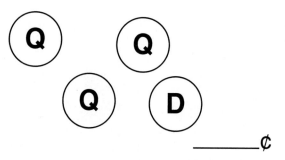

_____ ¢

11.

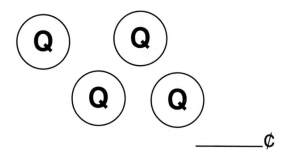

_____ ¢

Challenge

What 4 coins will make each amount?

12. 61¢

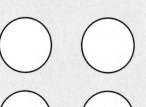

13. 80¢

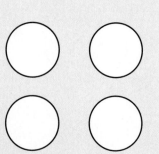

Problem Solving Strategy
Look for a Pattern

NCTM Standards 1, 2, 3, 6, 7, 8, 9, 10

🔻 **TEKS 1.4, 1.11B**

Understand

Plan

Solve

Check

Name _____ Date _____

1. Kylie is older than her brother, Ryan.

The table shows their ages at different times.

Kylie's Age	3	4	5	6	7
Ryan's Age	1	2	3	4	5

How old will Ryan be when Kylie is 10 years old?

_____ years old

2. Ben builds a staircase with blocks.
This staircase is 5 steps tall.

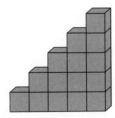

How many more blocks will he need to add another step?

_____ more blocks

3. Kira made a design with square tiles.

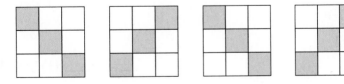

What will the next part of her design look like?

NOTE: Your child is exploring different ways to solve
problems. Looking for a pattern can help you
see how information in the problem is related.

Problem Solving Test Prep

1. Jenna made a tower with 4 blocks.

- The red block is above the yellow block.
- The green block is above the red block.
- The blue block is below the yellow block.

Which block is on the bottom?

Ⓐ blue   Ⓒ red

Ⓑ green   Ⓓ yellow

2. Darcy's bowling ball hit some of the 10 pins. She hit 2 more pins than the number of pins still standing.

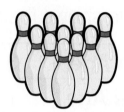

How many pins did she knock down?

Ⓐ 2   Ⓒ 8

Ⓑ 6   Ⓓ 10

 Show What You Know

3. Sammy had 2 dimes, 3 nickels, and 2 pennies. He spent 12¢.

Which coins could he have left?

_____

_____

Use words, numbers, or pictures to explain.

_____

_____

4. Geri sorts these figures. How many more figures have 4 sides than 3 sides?

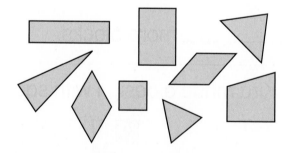

_____ more figures

Write a number sentence to show how you found your answer.

_____

Review/Assessment

NCTM Standards 1, 2, 6, 7, 8, 9, 10

1. Draw the missing jumps. Complete the table. Lessons 1–2

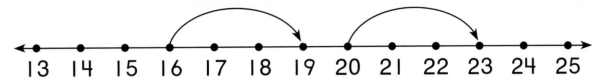

Start	16	20	13		22
Land	19	23		18	

What is the missing number? Lessons 3–4

2. $\boxed{31} + \boxed{10} = \boxed{}$

3. $\boxed{31} - \boxed{10} = \boxed{}$

4. $\boxed{27} + \boxed{} = \boxed{37}$

5. $\boxed{27} - \boxed{} = \boxed{17}$

---

6. What numbers are missing? Lesson 5

Start	13	26	34	47		19	
Jump Forward	2	4	5		3	0	10
Land	15	30		48	25		40

What is the number? Lessons 6 and 8

7. 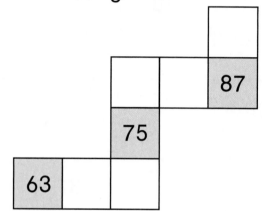 _____

8. _____

9. eighteen _____

10. fifty-seven _____

11. twenty-four _____

12. thirty-one _____

13. What numbers are missing? Lesson 7

			87
	75	
63		

What is the value? Lesson 9

14.  _____ ¢

15.  _____ ¢

Problem Solving Lesson 10

16. Jason made a design with blocks. How many blocks will he need to make the next shape?

_____ blocks

Name _____

Doubling, Halving, and Fractions
Fair Shares

Take a handful of counters out of the bag.

STEP 1 Counting

How many counters did you take? _____
Do you think you can share the
counters equally with your partner? _____

STEP 2 Sharing

Are there leftover counters? _____

Is the total number of counters even or odd? _____

STEP 3 Predicting Leftovers

Repeat the activity several times. Record each time.

Number of Counters					
Any Leftover?					
Even or Odd?					

How can you predict if there will be a leftover?

_____

_____

_____

Investigation

School-Home Connection

Dear Family,

Today we started Chapter 8 in *Think Math!* In this chapter, I will learn about mirror lines and I will draw mirror images. I will double and halve amounts of money and find a number that is halfway between two numbers. I will also learn how to write halves, fourths, and thirds. There are NOTES on the Lesson Activity Book pages to explain what I am learning every day.

Here are some activities for us to do together at home. These activities will help me understand doubling, halving, and fractions.

Love,

_____

Family Fun

Doubling Coins

Work with your child to double a collection of dimes, nickels, and pennies.

- Gather 18 pennies, 18 nickels, 18 dimes and a number cube.

- Players take turns tossing the number cube and taking that number of coins from a mixed pile of dimes, nickels, and pennies.

- Each player counts to find out how much the collection is worth and then uses more coins to make a collection that is worth twice as much.

- Players use their amounts to complete this sentence: "I had _____¢. I doubled my coins and then I had _____¢."

21¢   42¢

Halfway Between

Work with your child to find the number that is halfway between two numbers on a number line.

- Gather two number cubes and two small objects such as dry beans or toys.

- On a large sheet of paper draw a number line from 0 to 12 as shown below.

```
0  1  2  3  4  5  6  7  8  9  10  11  12
```

- Place one object above 0 on the number line. Toss the number cubes and find the sum. Place the other object above this number on the number line.

- Each player guesses the number that is halfway between these numbers. Then work together to move the objects toward each other, one step at a time.

- The player whose guess was the closest, wins a point. Play for at least four more rounds.

Doubling Your Money

NCTM Standards 1, 2, 6, 7, 8, 9, 10
TEKS 1.3A, 1.5D

What are the missing costs?

Use coins to help you.

1.	5¢	10¢
2.	10¢	
3.	6¢	
4.	8¢	
5.	20¢	
6.		60¢
7.		50¢
	△	△ + △

NOTE: Your child is learning to double collections of coins
and amounts of money. To practice, tell your child the cost of
one item and ask your child to tell you the cost of two items.

Complete each sentence.
You may use coins to help.

8. If one balloon costs 50¢,

then two balloons cost _____.

---

9. If one balloon costs _____¢,
then two balloons cost 30¢.

---

10. If one balloon costs 25¢,

then two balloons cost _____¢.

---

11. If one balloon costs 20¢,

then two balloons cost _____¢.

Make your own.

12. If one balloon costs _____¢,

then two balloons cost _____¢.

Problem Solving

13. If a big balloon costs 24¢ and a little one costs 7¢, then two big balloons and two little balloons

cost _____¢.

Chapter 8
Lesson 3

Sharing to Find Half

NCTM Standards 1, 2, 6, 7, 8, 9

🔻 **TEKS 1.1C, 1.2A**

What are the missing costs?

Use coins to help you.

	Total 🔺	Half ◺
1.	20¢	10¢
2.	40¢	
3.		30¢
4.	50¢	
5.		15¢
6.	18¢	
7.	24¢	

NOTE: Your child is learning to halve money amounts. Ask your child to share a collection of dimes and pennies equally between two people.

Complete each sentence.

8. Two friends share 80 marbles equally.

Each friend gets ___40___ marbles.

9. Two friends share _____ marbles equally.
Each friend gets 24 marbles.

10. Two friends share 26 marbles equally.

Each friend gets _____ marbles.

Make your own.

11. Two friends share _____ marbles equally.

Each friend gets _____ marbles.

Problem Solving

12. A costs 62¢.

A ⟍ costs half as much.

A and a ⟍ cost _____¢.

Name _____   Date _____

Exploring One Half

NCTM Standards 1, 2, 3, 6, 7, 8, 9, 10

TEKS 1.2A, 1.2B

Is one half shaded?

1.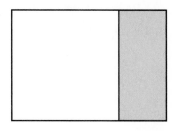

yes
(no)

2.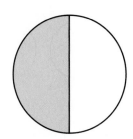

yes
no

3.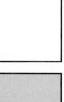

yes
no

4.

yes
no

5.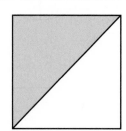

yes
no

6.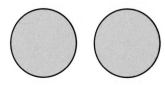

yes
no

How much is shaded?

7.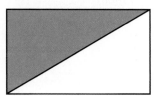

_____

8.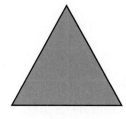

_____

9.

_____

NOTE: Your child is learning to find half of an object and half of a set of objects. Ask your child to find half of a collection of objects in your home.

Color one half.

10.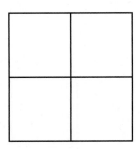

11.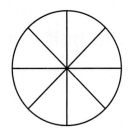

12.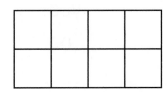

13.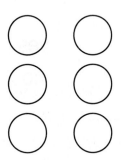

14.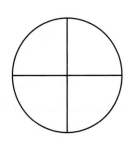

15.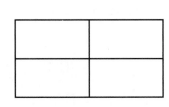

16.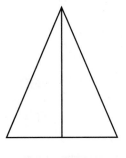

17.

18.

Problem Solving

19. Talia wants to buy a pack of stickers.
The stickers costs 60¢.
The store is having a half-off sale.
What is the sale price?

_____¢

Name _____ Date _____

Wholes and Halves

NCTM Standards 1, 2, 6, 7, 8, 9, 10

1. Match the pictures to the numbers.

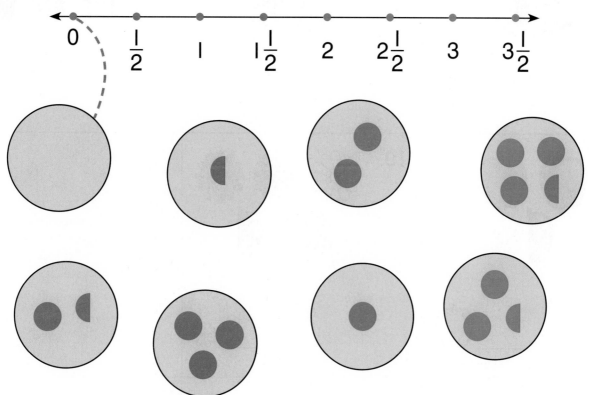

How many circles are there?

2.

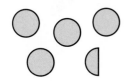

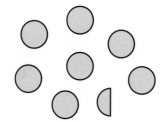

3.

_____

4.

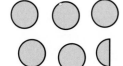

_____

5.

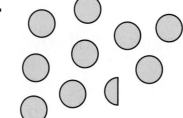

_____

NOTE: Your child is beginning to identify quantities and write numbers that combine wholes and halves. Ask your child to tell you how many wholes and how many halves are in each picture on this page.

How many are there?

6.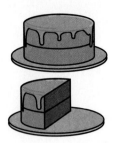

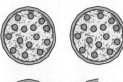

_____

7.

_____

8.

_____

9.

_____

10.

_____

11.

_____

12.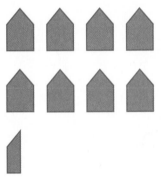

_____

13.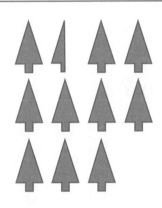

_____

14.

_____

Challenge

15. What number is halfway between 0 and 5?

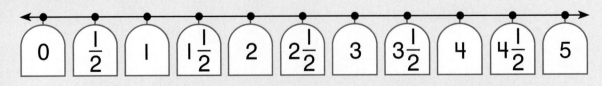

| 0 | $\frac{1}{2}$ | 1 | $1\frac{1}{2}$ | 2 | $2\frac{1}{2}$ | 3 | $3\frac{1}{2}$ | 4 | $4\frac{1}{2}$ | 5 |

Halfway Between Whole Numbers

NCTM Standards 1, 2, 6, 8, 9, 10

TEKS 1.5A

Complete the number line.
Then circle the number in the middle.

1.
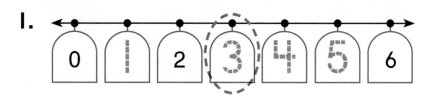
0 1 2 3 4 5 6

2.
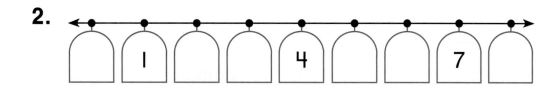
1   4   7

3.
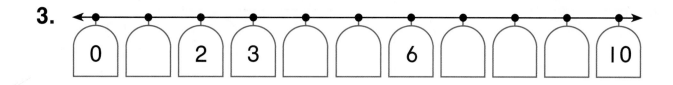
0   2 3   6   10

4.
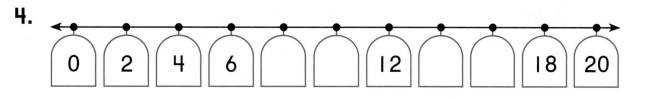
0 2 4 6   12   18 20

5. How did you find the numbers in the middle?

_____

_____

_____

_____

NOTE: Your child is learning to use number lines to find half of even and odd numbers. Have your child find the number that is halfway between 0 and 5.

What number is in the middle?

Use the number line to help you.

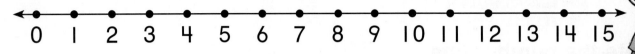

0 1 2 3 4 5 6 7 8 9 10 11 12 13 14 15

6.

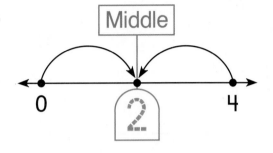

Middle

0   2   4

7.

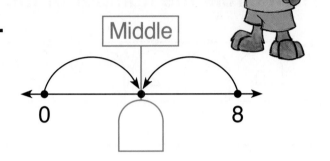

Middle

0   8

8.

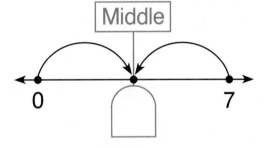

Middle

0   7

9.

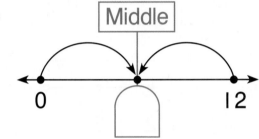

Middle

0   12

10.

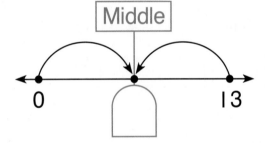

Middle

0   13

11.

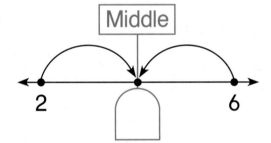

Middle

2   6

Challenge

12. Write the missing numbers.

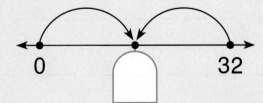

0   32

☐ + ☐ = 32

Chapter 8
Lesson 7

Half of a Half

NCTM Standards 1, 6, 7, 8, 9, 10
 TEKS 1.2A, 1.2B

Is one fourth blue?

1.
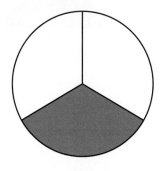
yes
(no)

2.
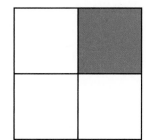
yes
no

3.

yes
no

4.
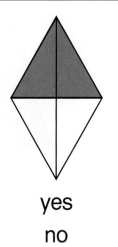
yes
no

5.
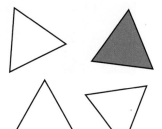
yes
no

6.

yes
no

How much is blue?

7.

_____ out of _____
equal parts

8.
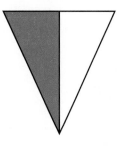
_____ out of _____
equal parts

9.
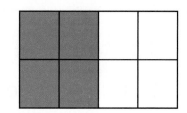
_____ out of _____
equal parts

 NOTE: Your child is learning to identify fourths.
Ask your child to identify what part of each
figure or set of figures is not shaded.

© Education Development Center, Inc.

Write a fraction for the green part.

10.

$\dfrac{3}{4}$

_____

11.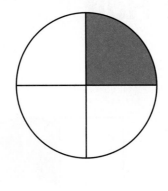

_____

12.

_____

13.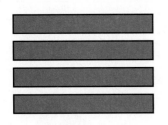

_____

14.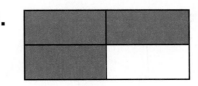

_____

15.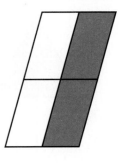

_____

Problem Solving

16. Geri wants to give $\dfrac{1}{4}$ of her stickers to her brother. She has 8 stickers. How many stickers will her brother get?

_____ stickers

Name _____ Date _____

Thirds

NCTM Standards 1, 6, 7, 8, 9, 10

 TEKS 1.2A, 1.2B

Is one third blue?

1.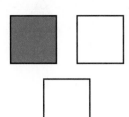

(yes)   no

2.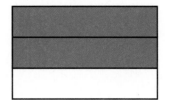

yes   no

3.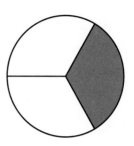

yes   no

4.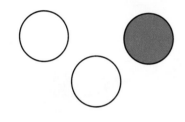

yes   no

5.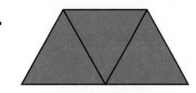

yes   no

6.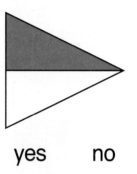

yes   no

How much is blue?

7.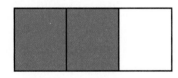

2 out of 3
equal parts

$\frac{2}{3}$

8.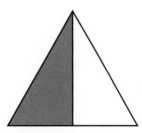

____ out of ____
equal parts

____

9.

____ out of ____
equal parts

____

© Education Development Center, Inc.

NOTE: Your child is learning to identify and write thirds.
Ask your child to identify the parts of Problems 1 to 6
that are not shaded.

Write a fraction for the green part.

10.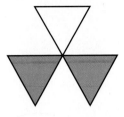

$\frac{2}{3}$

11.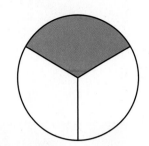

___

12.

___

13.

___

14.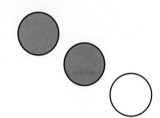

___

15.

___

Challenge
Write < or >.

16. $\frac{1}{3}$ ◯ $\frac{2}{3}$

17. $\frac{1}{2}$ ◯ $\frac{1}{3}$

Name _____ Date _____

Problem Solving Strategy
Guess and Check

NCTM Standards 1, 4, 6, 7, 8, 9, 10

🔻 TEKS 1.11B, 1.11C

Understand
Plan
Solve
Check

1. The school fair starts at 10:30. It ends at 3:30.
What time will it be halfway through the fair?

Start
10:30

End
3:30

_____

---

2. Two pencils cost 26¢.
Don buys 1 pencil.
How much does the pencil cost?

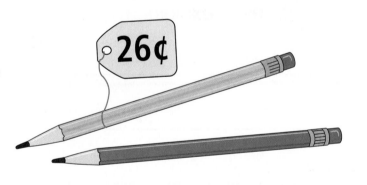

_____¢

---

3. Jack has some marbles.
Ben has twice as many.
Together, they have 12 marbles.
How many marbles does each
boy have?

Jack has _____ marbles.

Ben has _____ marbles.

🏠 **NOTE:** Your child is exploring different ways to solve problems.
Guessing and checking can help you get started. Using your first
guess can help to make a better second guess.

Problem Solving Test Prep

1. Thelma has nickels and pennies in her piggy bank. How many ways can she make 10¢?

    (A) 2

    (C) 4

    (B) 3

    (D) 10

2. Carl has 30¢ today. Yesterday he lost 2 dimes. How much money did Carl have before?

    (A) 10¢

    (C) 40¢

    (B) 30¢

    (D) 50¢

✏ Show What You Know

3. Chris had 10¢. Then he found 3 nickels. He bought an eraser for a dime. How much money does he have now?

    _____

    Use words, numbers, or pictures to explain.

    _____

    _____

    _____

    _____

    _____

    _____

4. Dori is in front of Keith. Sam is behind Dori. There are no others in line. Who is at the front of the line?

    _____

    Explain how you know your answer is correct.

    _____

    _____

    _____

    _____

    _____

    _____

Chapter 8 **Review/Assessment**

NCTM Standards 1, 2, 3, 4, 6, 7, 8, 9, 10

Draw a line to show halves. Lesson 1

1.

2.

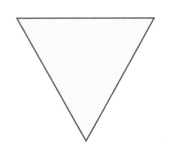

3.

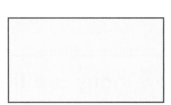

Draw the mirror image. Lesson 1

4.

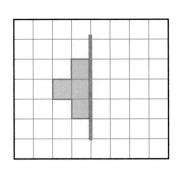

5.

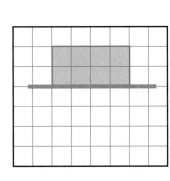

6.

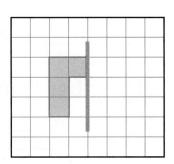

Complete each sentence. Lessons 2 and 3

7. If one book costs 25¢,

then two books cost _____¢.

8. Two friends share 40 pennies equally.

Each friend gets _____ pennies.

Write a fraction for the shaded part. Lessons 4, 7, and 8

9.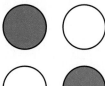

$\frac{2}{4}$

10.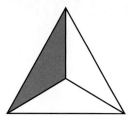

_____

11.

_____

How many are there? Lesson 5

12.

_____

13.

_____ 🍞

14.

_____ ○

What number is in the middle? Lesson 6

15.

Middle

0   12

16.

Middle

0   9

Problem Solving Lesson 9

17. The school play begins at 1:30 and ends at 3:30.

What time will it be halfway through the play?

_____

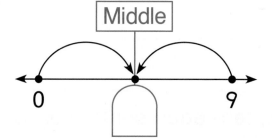
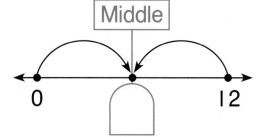

Name _____

Modeling Addition and Subtraction

Adding and Subtracting with Two Colors ✏️

➡️ TEKS 1.3B, 1.5E

You need
- 10 two-color counters, paper cup

Put all the counters in the cup.

STEP 1 **Spilling the Counters**

How many counters are red? _____

How many counters are yellow? _____

STEP 2 **Writing Addition Sentences**

Write an addition sentence to match the counters.

_____

Is there another addition sentence you can write? Explain.

_____

_____

STEP 3 **Writing Subtraction Sentences**

Write a subtraction sentence to match the counters.

_____

Is there another subtraction sentence you can write? Explain.

_____

_____

Investigation

Dear Family,

Today we started Chapter 9 of *Think Math!* In this chapter, I will add and subtract using Cuisenaire® Rods, rule machines, and Stair-Step Numbers. There are NOTES on some of the Lesson Activity Book pages to explain what I am learning every day.

Here are some activities for us to do together at home. These activities will help me practice addition and subtraction.

Love,

_____

Family Fun

Guess the Rule

Work with your child to play a game called *Guess the Rule.* Your child will play this game in Lesson 3.

- Think of a secret rule for the rule machine, such as add 4, subtract 3, 1 hour later, spend a nickel, or double. Tell your child if the rule involves time, money, or numbers.

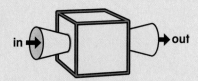

- Have your child say an input. Then you say the output. Continue until your child is ready to guess your rule. You may wish to record the numbers in an input/output table.

in	5	8	2	1	10
out	9	12	6	5	14

Fact Family Toss

Work with your child to practice writing addition and subtraction fact families.

- Have your child toss two number cubes labeled 1 through 6. Ask your child to find the sum of the numbers tossed.

- Then have your child write a fact family for these numbers.

  $$3 + 5 = 8 \quad 8 - 5 = 3$$
  $$5 + 3 = 8 \quad 8 - 3 = 5$$

- You might ask what the fact family will look like if two of the same numbers are tossed.

- Have your child repeat this activity a few more times for extra practice.

Chapter 9
Lesson 1

Exploring Addition with Cuisenaire® Rods

NCTM Standards 1, 2, 6, 9, 10

⬥ TEKS 1.3B

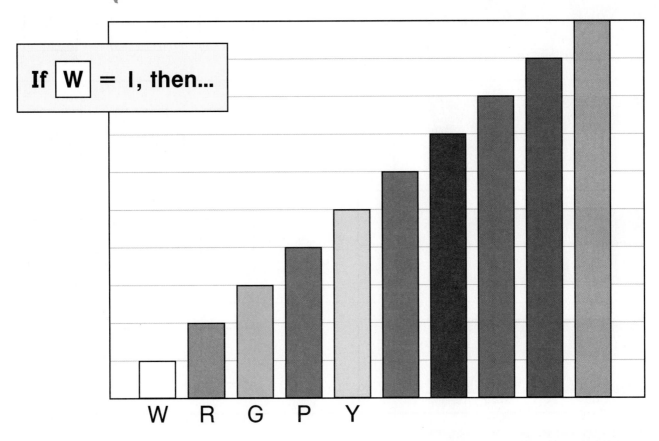

If W = 1, then...

W   R   G   P   Y

1. Complete the code.
Use a different letter for each rod.

W	R	G	P	Y					
1	2								

2. Complete the number sentence.

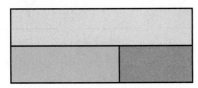

G + ___ = ___

[ ] + [ ] = 5

NOTE: Your child is learning to write addition sentences about Cuisenaire® Rods. Ask your child to write an addition sentence about any two rods.

Complete each addition sentence.

3. $1 + 5 = 6$

4. $\boxed{} + \boxed{} = 6$

5. $\boxed{} + \boxed{} = 6$

6. $\boxed{} + \boxed{} = 6$

7. $\boxed{} + \boxed{} = 6$

Challenge

8. What rod is missing? Complete the addition sentence.

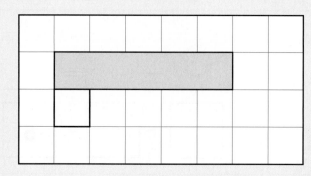

_____ + _____ = _____ Y

Recording Addition Sentences

NCTM Standards 1, 2, 6, 9, 10

TEKS 1.3B

Complete each addition sentence.

If $\boxed{W}$ = 1, then...

1.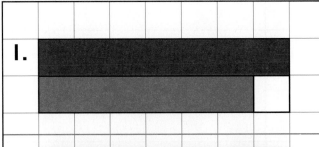

$6 + 1 = 7$

2.

$\square + \square = 7$

3.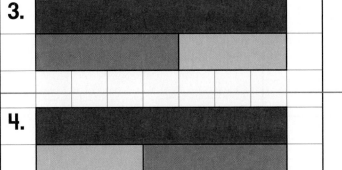

$\square + \square = 7$

4.

$\square + \square = 7$

5.

$\square + \square = 7$

6.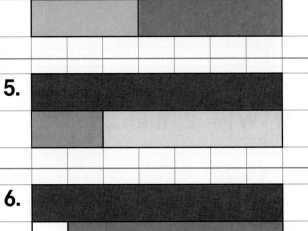

$\square + \square = 7$

NOTE: Your child is learning to find different ways to make the same number. Ask your child to find at least two ways to make a sum of 5.

Complete each addition sentence.

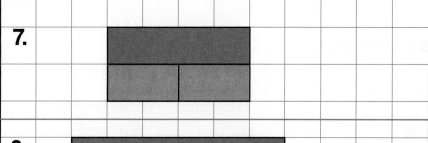

7.

$\boxed{2} + \boxed{2} = \boxed{4}$

8.

$\boxed{} + \boxed{} = \boxed{}$

9.

$\boxed{} + \boxed{} = \boxed{}$

10.

$\boxed{} + \boxed{} = \boxed{}$

Challenge

If $\boxed{W}$ = 2, then...

11.

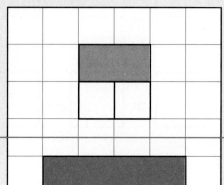

$\boxed{2} + \boxed{2} = \boxed{}$

12.

$\boxed{} + \boxed{} = \boxed{}$

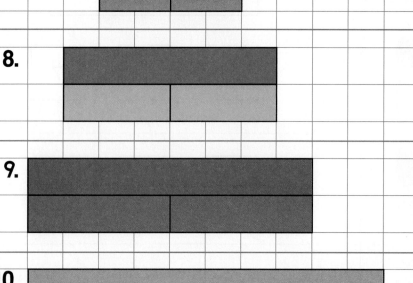

Name _____ Date _____

Exploring Input/Output Tables

NCTM Standards 1, 2, 4, 6, 9, 10
TEKS 1.3B, 1.8B, 1.11C

What is missing?

1.

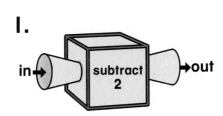

in	out
8	6
11	9
12	10
6	4
4	
	3
10	
	5

2.

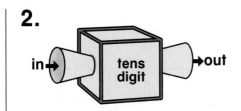

in	out
32	3
19	1
75	
65	
98	
32	
	5
	4

3.

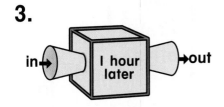

in	out
3:00	4:00
1:00	2:00
8:30	:
2:00	:
:	5:00
:	8:30
10:30	:
:	1:00

NOTE: The tables in this lesson help your child recognize patterns. You may wish to create similar rule machines to share with your child.

What is missing?

4.

in

first
letter

out

in	hog	ill	top	hat	ear	rug
out	h	┊				

5.

in

spend
10¢

out

in	40¢	50¢	60¢			35¢
out	30¢	40¢		60¢	5¢	

Challenge

6. Make your own rule.

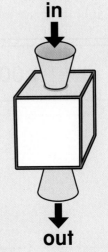

in

out

in				
out				

Name _____ Date _____

Using Input/Output Tables

NCTM Standards 1, 2, 6, 9, 10

TEKS 1.11C

What is missing?

1.

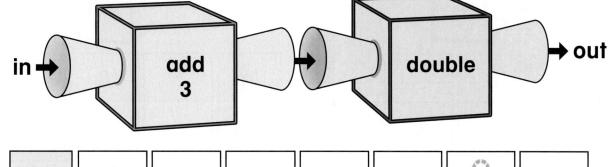

in	3	1	7	4	5	0	
out	12	8				6	10

2.

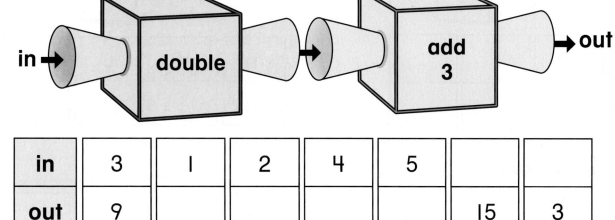

in	3	1	2	4	5		
out	9					15	3

3.

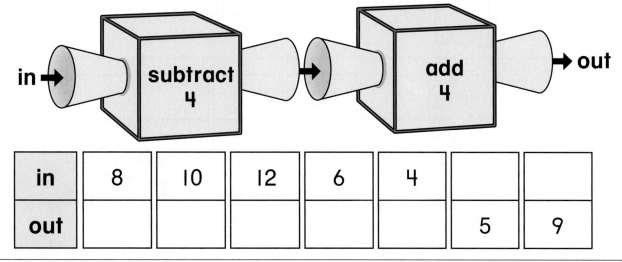

in	8	10	12	6	4		
out						5	9

NOTE: Your child is learning to complete multi-step problems. Ask your child to explain how to find the missing numbers in one of the tables.

What is missing?

4.

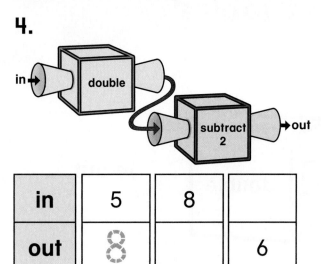

in	5	8	
out	8		6

5.

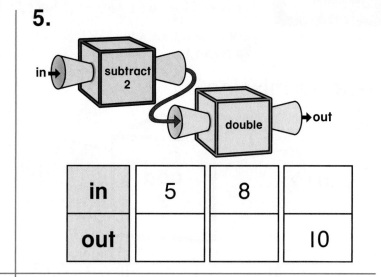

in	5	8	
out			10

6.

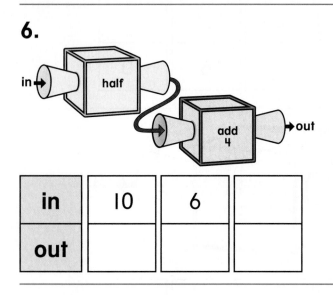

in	10	6	
out			

7.

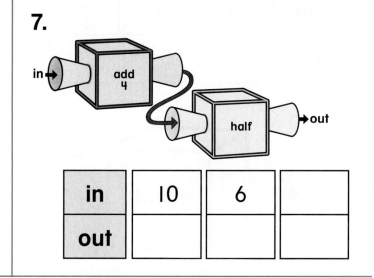

in	10	6	
out			

8. Make your own.

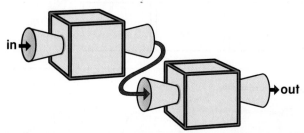

in	4	3	
out			

Challenge
9.

in	4	10	8		
out				9	5

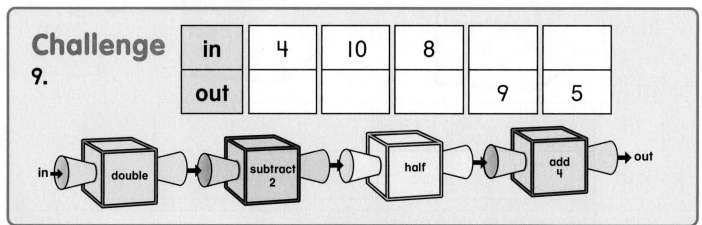

Chapter 9
Lesson 5 **Making Fact Families**

NCTM Standards 1, 2, 6, 9, 10
TEKS 1.3B, 1.5E

What is the fact family?

1.

$$1 \quad + \quad 8 \quad = \quad 9$$

$$8 \quad + \quad 1 \quad = \quad 9$$

$$9 \quad - \quad 1 \quad = \quad \underline{}$$

$$9 \quad - \quad \underline{} \quad = \quad 1$$

2.

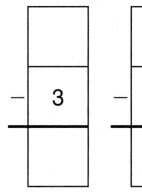

3.

$$4 \quad + \quad \underline{} = \underline{}$$

$$5 \quad + \quad \underline{} = \underline{}$$

$$\underline{} - \quad 4 \quad = \underline{}$$

$$\underline{} - \quad 5 \quad = \underline{}$$

NOTE: Your child is learning to write addition and subtraction fact families. Ask your child to explain how to complete one of the exercises.

What is the fact family?

4.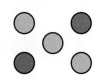

$$3 + 2 = 5$$

___ + ___ = ___

___ − ___ = ___

___ − ___ = ___

5.

___ + ___ = ___

___ + ___ = ___

___ − ___ = ___

___ − ___ = ___

6.

___ + ___ = ___

___ + ___ = ___

___ − ___ = ___

___ − ___ = ___

7. Color some dots. Write the fact family.

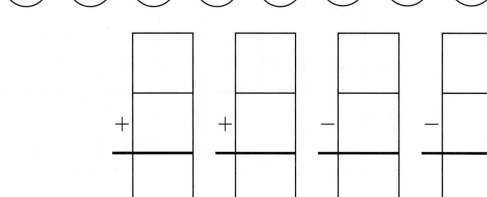

Problem Solving

8. There are 8 cups.
6 are empty.
The rest are full.
What is the fact family?

6 + ___ = 8

___ + ___ = 8

___ − ___ = ___

___ − ___ = ___

Fact Families and Stair-Step Numbers

NCTM Standards 1, 2, 6, 7, 8, 9, 10
TEKS 1.4, 1.5E

I. What is missing?

Draw dots to make the missing picture.

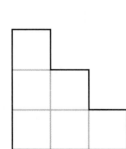
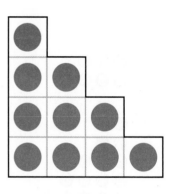

| 1 | 3 | | |

What is the fact family?

2.

3 + 1 =
_____

1 + 3 =
_____

4 − 1 =
_____

4 − 3 =
_____

3.

___ + ___ = ___

___ + ___ = ___

___ − ___ = ___

___ − ___ = ___

4.

___ + ___ = ___

___ + ___ = ___

___ − ___ = ___

___ − ___ = ___

NOTE: Your child is learning to write fact families using Stair-Step Numbers. Ask your child to explain one of the fact families on this page.

5. What is next?

6. Explain how you found the numbers in Problem 5.

_____

_____

_____

_____

What is the fact family?

7. 6 + _____ = 10

_____ + _____ = _____

_____ − _____ = _____

10 − 6 = _____

8. _____ + 5 = _____

_____ + _____ = _____

15 − 5 = _____

_____ − _____ = _____

Challenge

9. Look at Problem 5.
What are the next 3 Stair-Step Numbers? Use counters or draw a picture to help you.

Chapter 9
Lesson 7

Connecting Stories and Fact Families

NCTM Standards 1, 2, 6, 8, 9, 10

TEKS 1.5E

What is the fact family?

1. I have 5 pets.
2 are dogs.
The others
are kittens.

$$2 + 3 = 5$$
$$3 + 2 = 5$$
$$5 - 3 = \underline{\hphantom{0}}$$
$$5 - \underline{\hphantom{0}} = \underline{\hphantom{0}}$$

2. All the houses on our street
are blue or brown.
7 are blue.
3 are brown.

$$\underline{\hphantom{0}} + \underline{\hphantom{0}} = \underline{\hphantom{0}}$$
$$\underline{\hphantom{0}} + \underline{\hphantom{0}} = \underline{\hphantom{0}}$$
$$\underline{\hphantom{0}} - \underline{\hphantom{0}} = \underline{\hphantom{0}}$$
$$\underline{\hphantom{0}} - \underline{\hphantom{0}} = \underline{\hphantom{0}}$$

3. 6 of the marbles are blue.
The rest are green.
There are 11 marbles.

$$\underline{\hphantom{0}} + \underline{\hphantom{0}} = \underline{\hphantom{0}}$$
$$\underline{\hphantom{0}} + \underline{\hphantom{0}} = \underline{\hphantom{0}}$$
$$\underline{\hphantom{0}} - \underline{\hphantom{0}} = \underline{\hphantom{0}}$$
$$\underline{\hphantom{0}} - \underline{\hphantom{0}} = \underline{\hphantom{0}}$$

NOTE: Your child is learning to connect fact families to real-world situations. You may wish to create a story similar to the ones on this page and ask your child for the matching fact family.

© Education Development Center, Inc.

Complete each fact family.
Then write a story to match it.

4. $8 + \underline{\quad} = 12$ _____

       $\underline{\quad} + 8 = 12$ _____

       $12 - \underline{\quad} = 8$ _____

       $\underline{\quad} - \underline{\quad} = \underline{\quad}$ _____

5. $2 + 5 = \underline{\quad}$ _____

       $\underline{\quad} + \underline{\quad} = \underline{\quad}$ _____

       $\underline{\quad} - 5 = 2$ _____

       $\underline{\quad} - \underline{\quad} = \underline{\quad}$ _____

Problem Solving

6. Write a fact family that
    goes with the story. Solve.

    Amy has 6 books.
    Ken has 1 more book than Amy,
    Amy and Ken have

    _____ books

    $\underline{\quad} + \underline{\quad} = \underline{\quad}$

    $\underline{\quad} + \underline{\quad} = \underline{\quad}$

    $\underline{\quad} - \underline{\quad} = \underline{\quad}$

    $\underline{\quad} - \underline{\quad} = \underline{\quad}$

Name _____ Date _____

Two-Sentence Fact Families

NCTM Standards 1, 2, 6, 7, 8, 9, 10

⬥ TEKS 1.3B, 1.5E

What is the fact family?

I.

$$5 + 3 = 8$$

$$3 + 5 = 8$$

$$8 - 3 = 5$$

$$\_ - \_ = \_$$

---

2. There are 7 children.
2 wear jeans.
The rest wear shorts.

$$\_ + \_ = \_$$

$$\_ + \_ = \_$$

$$\_ - \_ = \_$$

$$\_ - \_ = \_$$

---

3. Bill and Kay each have the
same number of pennies.
Together they have 16 pennies.

$$\_ + \_ = \_$$

$$\_ + \_ = \_$$

$$\_ - \_ = \_$$

$$\_ - \_ = \_$$

🏠 **NOTE:** Your child is learning more about addition and
subtraction fact families. Ask your child to explain why
some fact families only have two number sentences.

Write the fact family. Then solve.

4.

_____ books

___ + ___ = ___

___ + ___ = ___

___ − ___ = ___

___ − ___ = ___

---

5. Sandra walks 4 blocks to school.
Then she walks home.
How many blocks did she walk?

_____ blocks

___ + ___ = ___

___ + ___ = ___

___ − ___ = ___

___ − ___ = ___

---

6. My birthday is two weeks
from today.
How many days is that?

_____ days

___ + ___ = ___

___ + ___ = ___

___ − ___ = ___

___ − ___ = ___

Challenge

7. Make a two-sentence fact family.

___ + ___ = 1

1 − ___ = ___

Name _____ Date _____

Problem Solving Strategy
Look for a Pattern ✏️
NCTM Standards 1, 2, 6, 7, 8, 9, 10
🔻 TEKS 1.11B, 1.11C

Understand
Plan
Solve
Check

1. Julio earns money every week. He saves money too. What rule does Julio follow?

Money Earned	$4	$8	$6	$10
Money Saved	$2	$4	$3	$5

_____

_____

2. Annie saves 5¢ every day. How much money will Annie save in a week?

_____ ¢

AUGUST						
Sunday	Monday	Tuesday	Wednesday	Thursday	Friday	Saturday

3. Chris plants red and yellow roses in his garden. He follows a pattern. What are the colors of the next two flowers?

_____

NOTE: Your child is exploring different ways to solve problems. Ask your child to explain how looking for a pattern can help solve these problems.

Problem Solving Test Prep

1. Trey plants beans and carrots. He has 10 plants. He has 2 more beans than carrots. How many carrots does he plant?

   (A) 2

   (B) 4

   (C) 6

   (D) 10

2. Carla made a pattern with counters.

   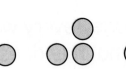

   How many counters are in the next figure?

   (A) 7

   (B) 9

   (C) 10

   (D) 12

 Show What You Know

3. You have these coins. You want to buy a sticker for 48¢. How much more money do you need?

   _____¢

   Explain how you found your answer.

   _____

   _____

   _____

4. Stickers cost 10¢ each. Joey has 34¢. What is the largest number of stickers he can buy?

   _____ stickers

   Explain how you found your answer.

   _____

   _____

   _____

   _____

Review/Assessment

NCTM Standards 1, 2, 6, 7, 8, 9, 10

I. Complete the addition sentences. Lessons 1 and 2

If | W | = I, then . . .

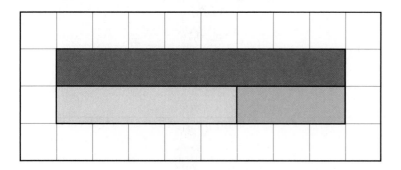

☐ + ☐ = **8**

What is missing? Lessons 3 and 4

2.

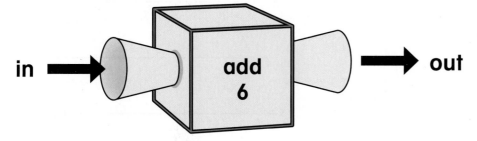

in ➡ | add 6 | ➡ out

in	2	5	6			9	7	
out				9	10			7

3.

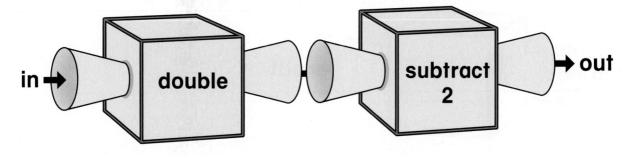

in ➡ | double | — | subtract 2 | ➡ out

in	3	2	6	7	4		
out	4					8	0

What is the fact family? _{Lessons 5–8}

4.

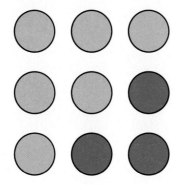

_____ + _____ = _____

_____ + _____ = _____

_____ − _____ = _____

_____ − _____ = _____

---

5. Jordan has 8 shirts.
Four are blue.
The rest are white.

_____ + _____ = _____

_____ + _____ = _____

_____ − _____ = _____

_____ − _____ = _____

Problem Solving _{Lesson 9}

6. Find a rule.
Complete the table.

in	4	6	3	9
out	8	12		18

in → ? → out

Rule: _____

_____

Name _____

Maps, Grids, and Geometric Figures
Lines and Letters

STEP 1 **Drawing Lines to Make Letters**

Draw two letters that have only straight lines.

First Letter _____   Second Letter _____

STEP 2 **Describing Lines**

How many lines did you use in your first letter?   _____
Describe which direction the lines go.

_____

_____

How many lines did you use in your second letter?   _____
Describe which direction the lines go.

_____

_____

STEP 3 **Drawing More Letters**

Draw other letters that have only straight lines.
Draw as many as you can.

_____

_____

_____

Dear Family,

Today we started Chapter 10 of *Think Math!* In this chapter, I will learn to give and follow directions on a map, to draw and recognize congruent figures, and to identify and compare two- and three-dimensional figures. There are NOTES on some of my pages to explain what I am learning every day.

Here are some activities for us to do together at home. These activities will help me understand maps, grids, and geometric figures.

Love,

_____

Family Fun

Grid Treasure Hunt

Play this game with your child.

- You and your child each need a 6-by-6 grid with one-inch boxes. Draw a large dot in the middle of each grid. The grids will be used as maps.

- Hide a "treasure" by secretly making an *X* somewhere on your map.

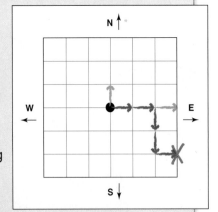

- Your child tries to find the location of the treasure by drawing an arrow one box long from the center dot on his or her map. If your child is getting farther away from the treasure, say "colder." If your child is getting closer, say "warmer."

- Your child continues to draw arrows until the location of the treasure is found.

Shape Search

Work with your child to identify familiar three-dimensional figures in your environment.

- On a piece of paper, make a chart with rows for each of the following figures: sphere (ball), cylinder (can), rectangular prism (box), and cone.

- Work with your child to find a few examples of each figure in your home and record the names of the objects in the chart.

Figure	Object
sphere	
cylinder	
rectangular prism	
cone	

- Discuss the figures you found. Which were easiest to find? Which were hardest to find? Which was the largest example of each figure? Which was the smallest?

Exploring Lines and Intersections

NCTM Standards 1, 3, 6, 8, 9, 10

TEKS 1.12A

Do the lines intersect?

1.

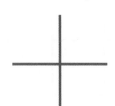

(yes)   no

2.

yes   no

3.

yes   no

4.

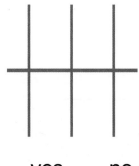

yes   no

5.

yes   no

6.

yes   no

Draw dots to show where the lines intersect.

7.

8.

9.

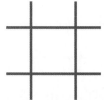

NOTE: Your child is learning to identify horizontal and vertical lines. Ask your child to identify which lines are horizontal, which are vertical, and where they intersect.

© Education Development Center, Inc.

Complete each table.

10.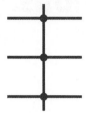

11.

12.

13.

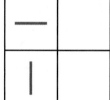

Challenge

14. Draw lines that intersect to match the table.

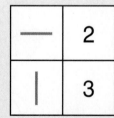

—	2
		3

Name _____ Date _____

Drawing Lines and Intersections

NCTM Standards 1, 3, 6, 8, 9, 10

Where do the lines intersect?
Complete each table.

1.

—	1
ǀ	2
+	2

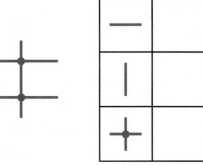

2.

—	
ǀ	
+	

3.

—	
ǀ	0
+	

4.

—	0
ǀ	
+	

5.

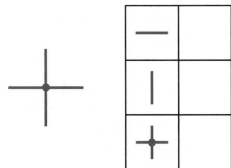

—	
ǀ	
+	

6.

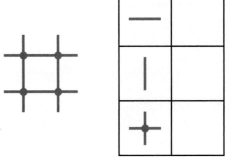

—	
ǀ	
+	

NOTE: Your child is recording information about horizontal and vertical lines, and the number of places they intersect. Ask your child to explain how to complete each table.

Complete each map. Write the missing numbers.

7.

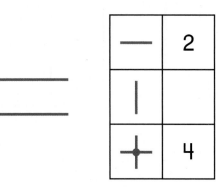

—	I
|	
+	

8.

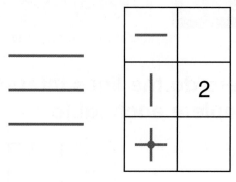

—	
|	2
+	

9.

—	3
|	3
+	

10.

—	2
|	3
+	

11.

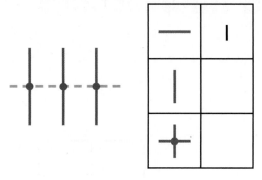

—	2
|	
+	4

12.

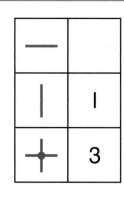

—	
|	I
+	3

Challenge

13. Draw the map and write the numbers.

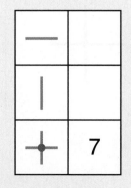

—	
|	
+	7

Name _____ Date _____

Exploring Direction on a Map

NCTM Standards 3, 6, 7, 8, 9, 10

🔻 **TEKS 1.13**

Find two paths from A to B on each map.

North ↑

1.

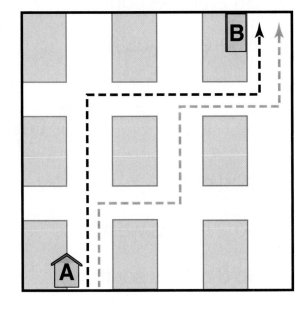

2.

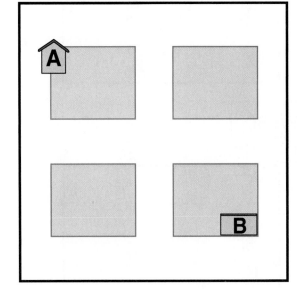

West
←

East
→

3.

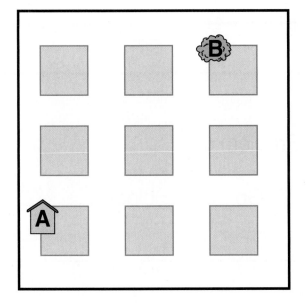

4.

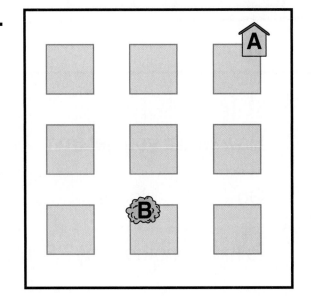

South ↓

© Education Development Center, Inc.

NOTE: Your child is learning to give and follow directions. Ask your child to explain the directions for each path on this page.

Follow the directions.

North ↑

5.

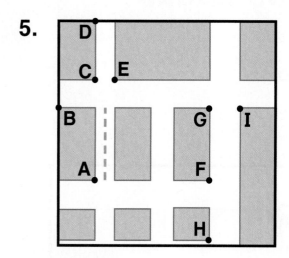

① Start at **A.**

② Go 1 block north.

③ Go 2 blocks east.

④ Go 2 blocks south.

⑤ Where are you now?

_____

West ←

East →

6.

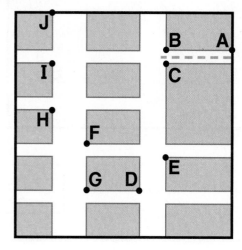

① Start at **A.**

② Go 2 blocks west.

③ Go 3 blocks south.

④ Go 1 block east.

⑤ Where are you now?

South ↓

_____

Problem Solving

7. How would you get from **J** to **E** in Problem 6?

① Go _____ blocks _____.   ② Go _____ block _____.

Is there another way? Explain: _____

_____

_____

Finding and Following Paths on a Grid

NCTM Standards 3, 6, 9, 10

Start at the dot. Draw each path.

1.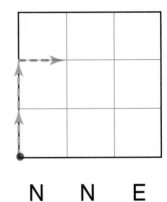

N   N   E

2.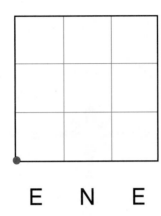

E   N   E

3.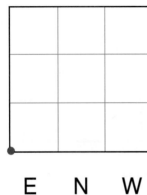

E   N   W

4.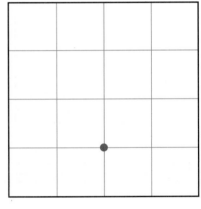

E   S   W   W

5.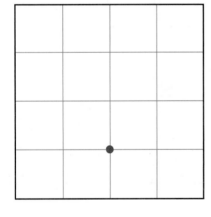

N   N   E   N

6.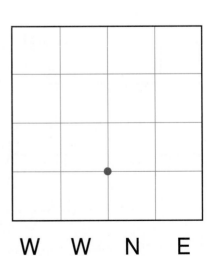

W   W   N   E

7.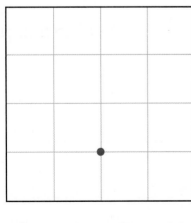

S   E   E   N

NOTE: Your child is learning to give and follow directions on a map. Ask your child to trace each path and say the directions.

8. Find different paths from **S** to **T.**

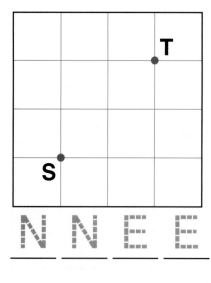

N N E E
___ ___ ___ ___

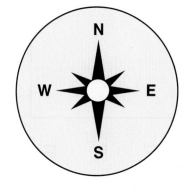

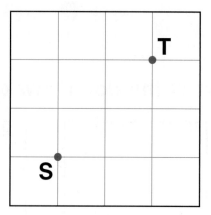

___ ___ ___ ___

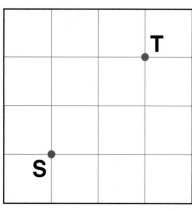

___ ___ ___ ___

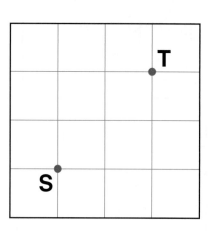

___ ___ ___ ___

Challenge

9. Find the shortest paths from **A** to **B.**

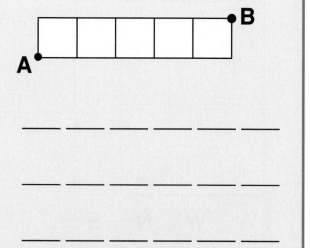

N E E E E E
___ ___ ___ ___ ___ ___

___ ___ ___ ___ ___ ___

___ ___ ___ ___ ___ ___

___ ___ ___ ___ ___ ___

___ ___ ___ ___ ___ ___

Paths and Figures on a Grid

NCTM Standards 3, 6, 9, 10

1. Start at the dot. Draw each path.

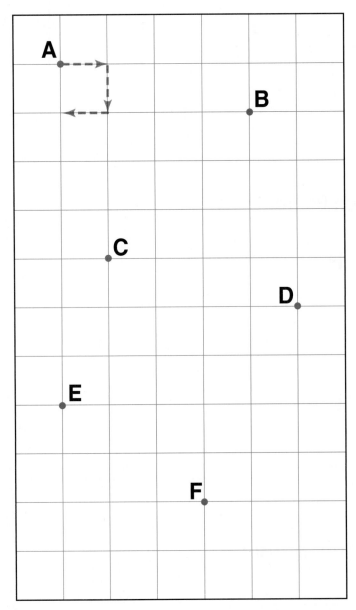

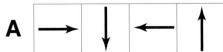

A

B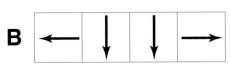

C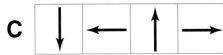

D

E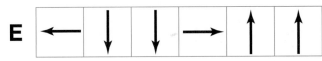

F

© Education Development Center, Inc.

NOTE: Your child is learning to give and follow directions
on a grid. Ask your child to tell you which paths on this page
start and end at the same point.

100 + 80 + 13   **CXCIII**   one hundred ninety-three   193

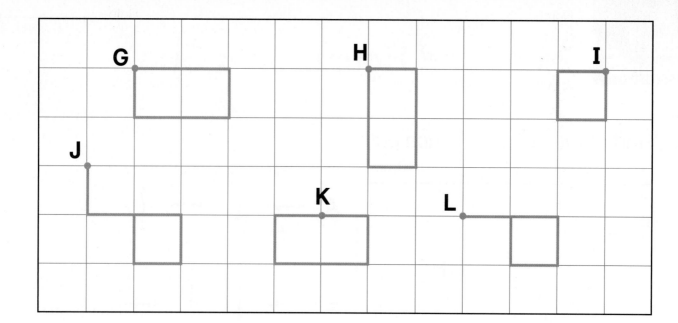

2. Write the directions for making each figure.

G

H

I

J

K

L

Look at the paths you drew on this page and on page 193.

Challenge
Look at the paths you drew on this page and on page 193.

3. Which paths start and end at the same point?

_____

4. Draw ↑, ↓, —→, or ←— to complete each sentence.

These paths use the same number of _____ as _____.

These paths use the same number of _____ as _____.

Name _____ Date _____

Exploring Symmetry

NCTM Standards 3, 6, 8, 9, 10

Start at the dot. Draw the mirror image.
Record the arrow directions.

1.

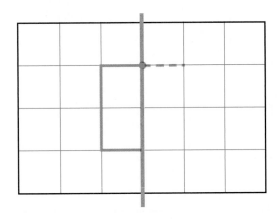

2.

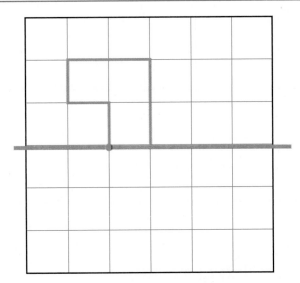

© Education Development Center, Inc.

NOTE: Your child is learning to recognize and draw symmetric figures. Have your child trace each figure and tell the directions to move on the grid.

Start at the dot. Draw the figure.
Then draw the mirror image. Write the directions.

3.

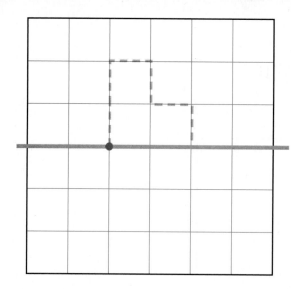

4.

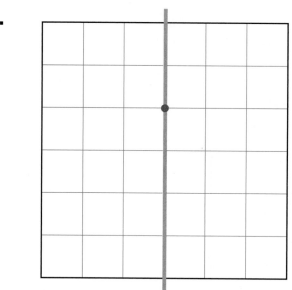

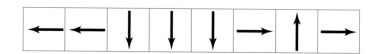

Challenge

5. Place a mirror along the gray line.
Circle the letters that show half in the mirror.

H T F O M I

Name _____ Date _____

Connecting Points to Make Figures

NCTM Standards 3, 6, 9, 10

🔸 TEKS 1.6A, 1.12B

A

B

D

C

1. Draw lines to connect the points.

A and **B**	**C** and **D**	**A** and **C**
B and **C**	**D** and **A**	**B** and **D**

 NOTE: Your child is learning to draw lines
to connect points. Ask your child to identify
the shapes of the figures that the lines make.

Draw lines to connect the points.

2.

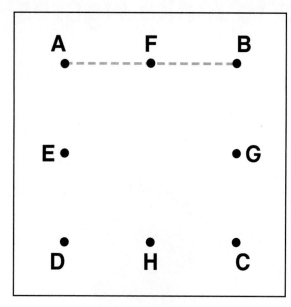

A and B   B and C
C and D   D and A
E and G   F and H
How many squares are there?

_____ squares

3.

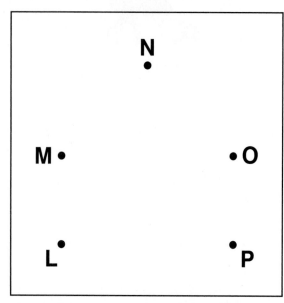

M and N   L and M
N and O   M and O
O and P   L and O
L and P
How many triangles are there?

_____ triangles

Problem Solving

4. Which 3 points can you connect to make a triangle?

Q   U   R

X   V

S   W   T

_____

_____

_____

Investigating Rectangles

NCTM Standards 1, 3, 4, 6, 7, 8, 9, 10
TEKS 1.6A, 1.12A

1. How many ☐ are in each figure?

Figure A

Figure B

Figure C

Figure D

Figure E

Figure F

2. Use the figures above to complete the table.

Figure	A	B	C	D	E	F
Number of ☐	4					

3. Is Figure F a rectangle? Explain.

_____

_____

NOTE: Your child is learning to identify and compare rectangles.
Have your child tell you the number of grid squares in each of the
rows and in each of the columns of the rectangles on this page.

4. How many ☐ are in each figure?

Figure G

`12`

Figure H

Figure I

Figure J

Figure K

Figure L

---

5. Use the figures above to complete the table.

Figure	G	H	I	J	K	L
Rows	1					
Columns	12					
Number of ☐	12					

Name _____   Date _____

Recording and Graphing Rectangles

NCTM Standards 1, 3, 4, 6, 9, 10

TEKS 1.12A

1. How many ☐ are in each figure?

Figure A	Figure B	Figure C
2	3	

Figure D	Figure E	Figure F

Figure G	Figure H	Figure I

2. Use the figures above to complete the table.

Figure	A	B	C	D	E	F	G	H	I
Number of ☐	2	3							

NOTE: Your child counted the number of grid squares in each rectangle and recorded the information in a table.

3. How many ☐ are in each figure?

Figure J
4

Figure K
9

Figure L

Figure M

Figure N

Figure O

---

4. Use the figures above to complete the table.

Figure	J	K	L	M	N	O
Rows	2					
Columns	2					
Number of ☐						

Finding Congruent Figures

NCTM Standards 3, 4, 6, 8, 9, 10

🔻 **TEKS 1.7D, 1.12A, 1.13**

Which figure does not have the same size and shape?

1.

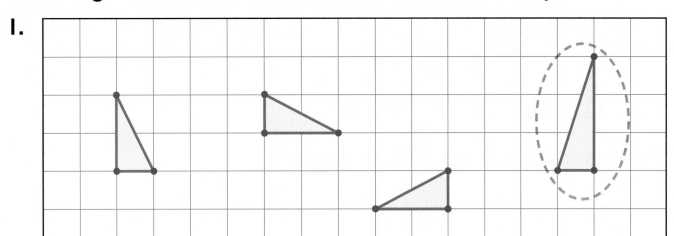

2.

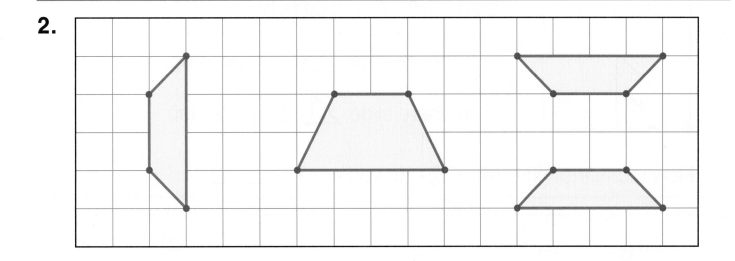

3.

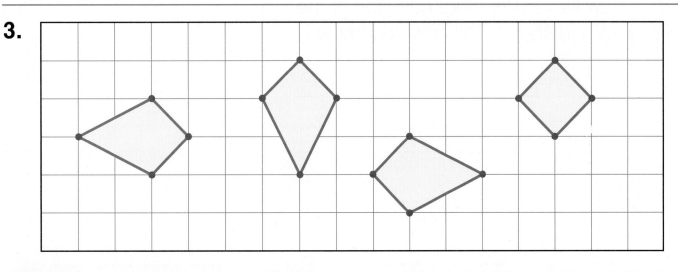

NOTE: Your child is learning about congruent figures.
Have your child identify which figures have the same
shape and size.

Slide the triangle to make figures that are the same. Draw the figures.

4.

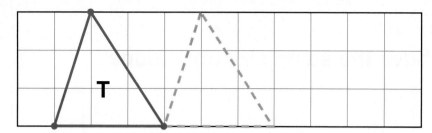

Slide 3 spaces right. Repeat.

How many triangles like do you see? _____

5.

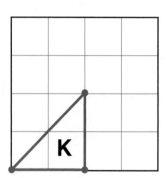

Slide 2 spaces right.

From there slide 2 spaces up.

From there slide 2 spaces left.

How many triangles like  do you see? _____

Problem Solving

6. Draw two figures that have the same size and shape. Tell how you know.

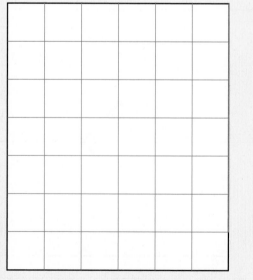

_____

_____

_____

Exploring
Three-Dimensional Figures

NCTM Standards 3, 6, 7, 8, 9, 10

⬇ **TEKS 1.6C, 1.12A**

I. Match the figures to their names
and to words that describe them.

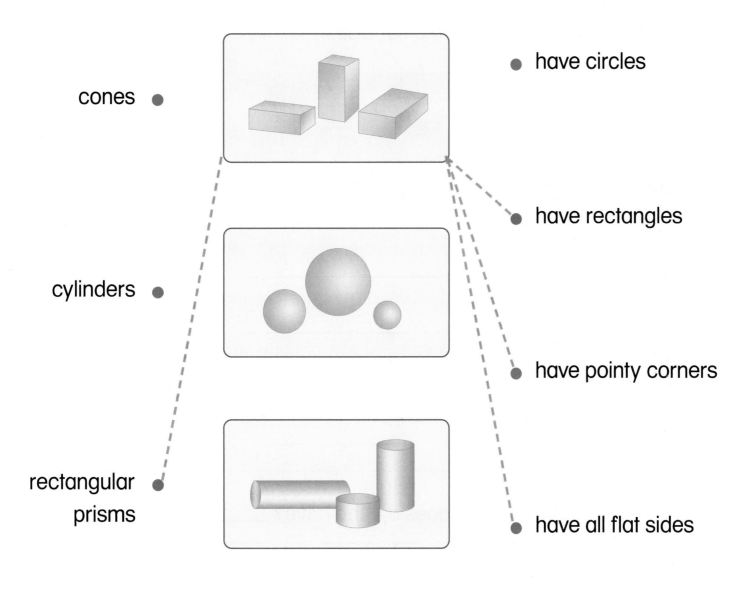

cones •

cylinders •

rectangular • prisms

spheres •

• have circles

• have rectangles

• have pointy corners

• have all flat sides

• have a curved side

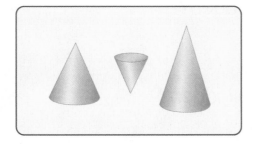

NOTE: Your child is learning to identify and compare three-dimensional figures. Ask your child to find examples of each of these figures in your home.

© Education Development Center, Inc.

2. How are a cone and a cylinder different?

_____

_____

_____

3. Circle the figures that have a flat side.

4. Give an example of something from the grocery store that has about the same shape.

sphere _____

cone _____

rectangular prism _____

cylinder _____

5. How many square faces does a cube have? _____

Challenge

6. How many faces are hidden?

_____ faces

7. How many corners are hidden?

_____ corner

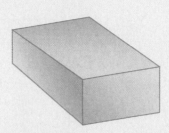

Name _____ Date _____

Problem Solving Strategy
Draw a Picture ✏

NCTM Standards 1, 2, 3, 5, 6, 7, 8, 9, 10

🔻 TEKS 1.10A, 1.11A, 1.11B, 1.11C, 1.12A

Understand
Plan
Solve
Check

1. How can you go from school to home?
Draw the path.
Write the directions.

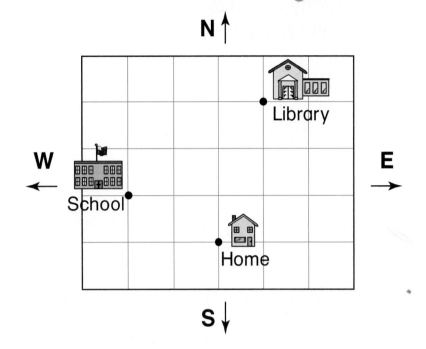

_____

_____

2. Use the arrows.
Will the figure show mirror images across the line? _____

left side

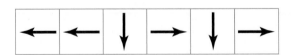

right side

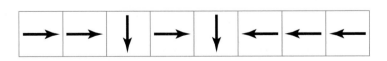

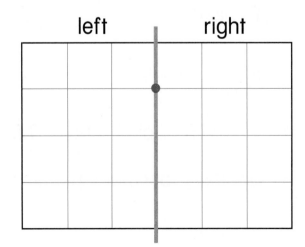

 NOTE: Your child is exploring different ways to solve problems. Sometimes drawing a picture is an efficient way to solve a problem.

Problem Solving Test Prep

1. There are 7 apples in a basket. 3 are red. The rest are green. Which number sentence tells how many green apples there are in the basket?

   (A) $7 + 3 = 10$

   (B) $7 - 3 = 4$

   (C) $10 - 3 = 7$

   (D) $10 + 7 = 17$

2. Eddie collects marbles. He buys 10 marbles every week. How many will he buy in 4 weeks?

   (A) 4 marbles

   (B) 8 marbles

   (C) 10 marbles

   (D) 40 marbles

 Show What You Know

3. How many more children chose bananas than apples? _____ children

Our Favorite Kinds of Fruit								
🍌	☺	☺	☺	☺	☺	☺		
🍎	☺	☺	☺	☺				

Key: Each ☺ stands for 1 child's choice.

Explain. _____

_____

4. I have 5 apples. I have 3 fewer pears. I have 2 more bananas than apples. How many pieces of fruit do I have? _____ pieces of fruit

Explain. _____

_____

Chapter 10 Review/Assessment

NCTM Standards 1, 3, 4, 6, 7, 8, 9, 10

Write the missing numbers. Lessons 1, 2

1.

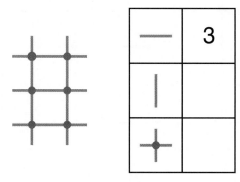

2.

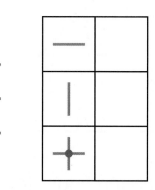

Follow the directions to draw each path.
Start at the dot. Lessons 3, 4

3. N E S S W N

4. W S W S E

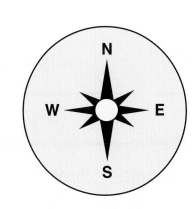

5. Start at the dot. Draw the mirror image. Record the directions. Lessons 5, 6

left   right

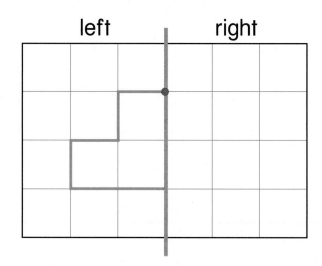

left side

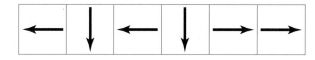

right side

6. Circle the figure that does not have the same size
and shape. Lesson 10

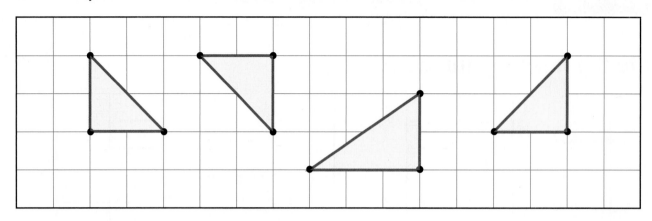

---

7. Match the figure to its name. Lesson 11

cylinder

rectangle

rectangular prism

Problem Solving Lessons 8, 9, 12

8. How many different rectangles cover 6 ☐?

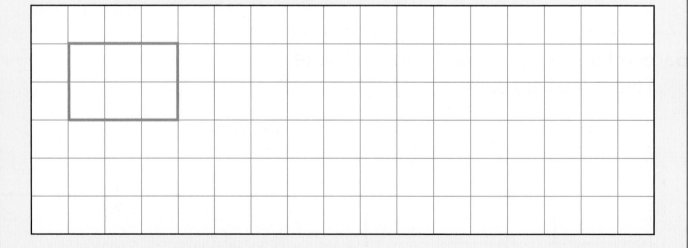

_____ rectangles

Comparing Numbers, Temperatures, and Weights
Numbers on a Thermometer ✏️

STEP 1 Estimating

How does the air feel inside?

_____

Take a look outside.
How do you think the air feels outside?

_____

STEP 2 Observing

Look at the numbers on a thermometer.
What pattern do you see?

_____

_____

STEP 3 Comparing

Is the temperature outside warmer than, cooler than, or about the same as the temperature inside?

_____

_____

How would the thermometer change if you brought it to a place that is cooler?

_____

Investigation

Dear Family,

Today we started Chapter 11 of *Think Math!* In this chapter, I wil learn to compare numbers, temperatures, and weights. I will learn to read a thermometer and to use a balance scale. There are NOTES on the Lesson Activity Book pages to explain what I am learning every day.

Here are some activities for us to do together at home. These activities will help me learn to compare and order numbers and to compare weights of objects.

Love,

_____

Family Fun

Larger and Smaller

Work with your child to identify numbers that are larger and smaller than a given number.

- Write a one- or two-digit number on a piece of paper. Display the number and have your child read it aloud.

- To the left of your number, have your child write a number that is smaller. To the right of your number, have your child write a number that is larger. Invite your child to explain how he or she chose the numbers.

- Repeat for other one- and two-digit numbers.

Heavy and Light

Work with your child to compare the weights of packaged household products.

- Have your child choose two packaged foods from your kitchen and feel the weight of each package.

- Ask your child to guess which package is heavier.

- Then point out the weights of the packages. Have your child tell which weight is greater. Then ask whether his or her guess was correct.

Name _____ Date _____

Comparing Groups

NCTM Standards 1, 2, 6, 7, 8, 9, 10
TEKS 1.1A, 1.12B

> is greater than
< is less than
= is equal to

Write <, >, or =.

1.

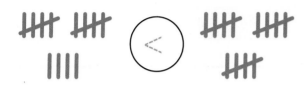

2.

3.

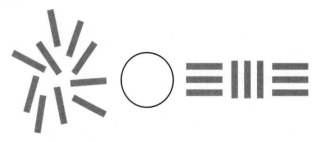

4.

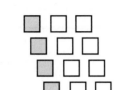

5.

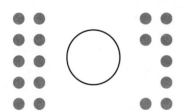

How many are there? Write <, >, or =.

6.

7.

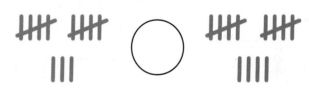

NOTE: Your child is learning to compare sets of objects. Ask your child to make two piles of pennies or paper clips and then to tell which pile has more.

© Education Development Center, Inc.

8. Write >, <, or =.

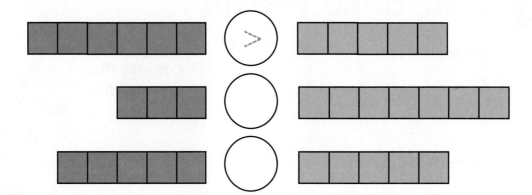

---

9. How can you use numbers and symbols to compare the strips in Problem 8?

_____

What is missing?

10. 3 < ☐ 4	**11.** 12 = ☐	**12.** 3 + 5 ◯ 1 + 5
13. 13 ◯ 13	**14.** 9 ◯ 7	**15.** 7 + 4 ◯ 4 + 7
16. ☐ < 7	**17.** ☐ > 10	**18.** 2 + 8 ◯ 2 + 5

Challenge

19. If ⬡ + ⬡ + ⬡ > 8

and ⬡ + ⬡ < 10 then

what could ⬡ be? _____

Comparing Numbers and Temperatures

NCTM Standards 1, 2, 4, 6, 7, 8, 9, 10
TEKS 1.1A, 1.7G, 1.12A

What numbers are missing?
Write <, >, or =.

1.

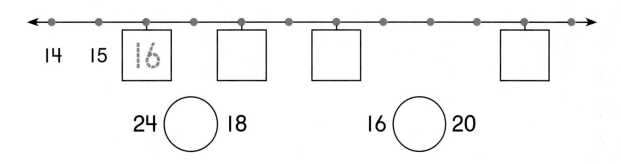

14   15   **16**   ☐   ☐   ☐

24 ◯ 18   16 ◯ 20

2.

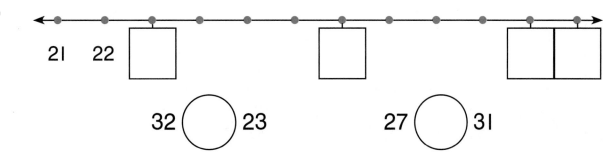

21   22   ☐   ☐   ☐ ☐

32 ◯ 23   27 ◯ 31

Write < or >.

3.

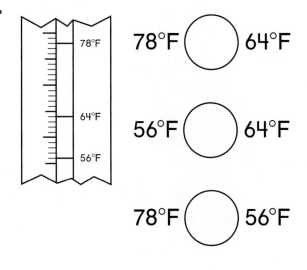

78°F ◯ 64°F

56°F ◯ 64°F

78°F ◯ 56°F

4.

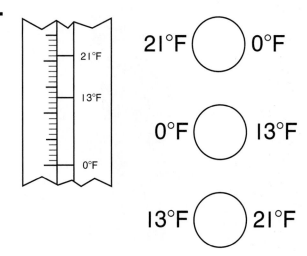

21°F ◯ 0°F

0°F ◯ 13°F

13°F ◯ 21°F

🏠 **NOTE:** Your child is comparing numbers on a number line and on a thermometer. Ask your child to explain how he or she chose which symbol to write in the problems on this page.

Color to show each temperature.
Circle the hotter temperature.

5.

$\widetilde{\left(60°F\right)}$ 40°F

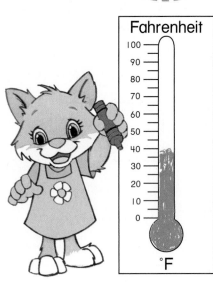

Fahrenheit
100
90
80
70
60
50
40
30
20
10
0
°F

Fahrenheit
100
90
80
70
60
50
40
30
20
10
0
°F

6.

75°F   85°F

Fahrenheit
100
90
80
70
60
50
40
30
20
10
0
°F

Fahrenheit
100
90
80
70
60
50
40
30
20
10
0
°F

Problem Solving

7. It is 55°F where Jacob lives.
It is 87°F where Ethan lives.
Which boy is Jacob and which is Ethan?

_____   _____

Name _____ Date _____

Using Place Value to Compare Numbers

NCTM Standards 1, 2, 6, 7, 8, 9, 10

TEKS 1.1A, 1.1B, 1.5C

What numbers are shown? Write <, >, or =.

1.

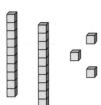 　　　　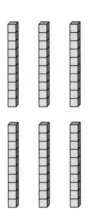

```
┌───────┐        ⬭        ┌───────┐
│  23   │                 │       │
└───────┘                 └───────┘
```

2.

```
┌───────┐        ⬭        ┌───────┐
│       │                 │       │
└───────┘                 └───────┘
```

3.

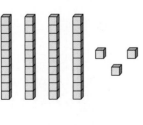 　　　

```
┌───────┐        ⬭        ┌───────┐
│       │                 │       │
└───────┘                 └───────┘
```

NOTE: Your child is learning to compare two-digit numbers. Ask your child to explain how he or she completed Problem 3 above.

4. Draw a picture to show each number.
Write <, >, or =.

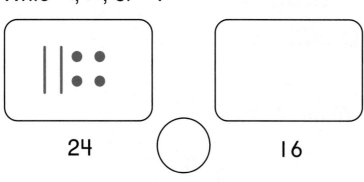

24 ◯ 16

You can draw | for rods and • for units.

Write <, >, or =.

5. 63 ⟩ 36

6. 45 ◯ 45

7. 18 ◯ 81

8. 28 ◯ 82

9. 93 ◯ 39

10. 68 ◯ 68

11. 8 ◯ 21

12. 79 ◯ 79

13. 19 ◯ 7

14. 25 ◯ 29

15. 51 ◯ 38

16. 32 ◯ 6

Challenge

17. What numbers are shown?
Write <, >, or =.

You can trade  for 10 ▭.

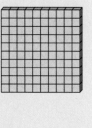 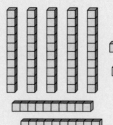 +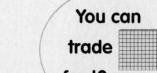

 ◯ +

Name _____ Date _____

Ordering Numbers

NCTM Standards 1, 2, 6, 7, 8, 9, 10
TEKS 1.1A, 1.5C, 1.12A, 1.12B, 1.13

Compare and order the numbers.
Write the missing symbols and numbers.

1.

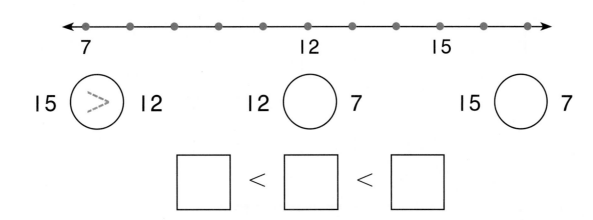

2.

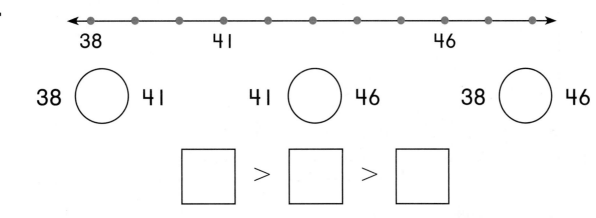

3.

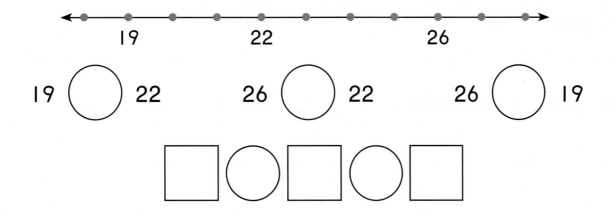

🏠 **NOTE:** Your child is continuing to compare two-digit numbers. This lesson introduces ordering three numbers from smallest to largest or from largest to smallest.

Write the numbers.
Then order the numbers.

4.

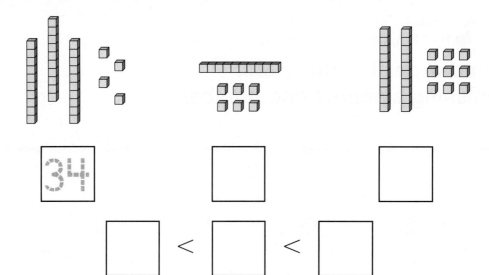

34   ☐   ☐

☐ < ☐ < ☐

5.

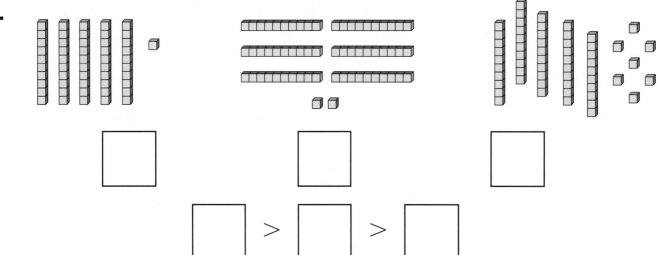

☐   ☐   ☐

☐ > ☐ > ☐

Problem Solving

6. There are more apples than bananas. So, a > b.
   There are fewer cherries than bananas. So c < b.
   Which is true?

   (A) a < c   (B) a > c   (C) a = c   (D) You can't tell.

   ✏ Explain how you know. _____

   _____

   _____

Name _____ Date _____

Changing Both Sides of a Sentence

NCTM Standards 1, 2, 6, 7, 8, 9, 10

TEKS 1.1A, 1.12A

Write <, >, or =.

1.
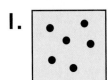

6 ⟩ 5   and   2 ◯ 2   so   6 − 2 ◯ 5 − 2

2.

3 ◯ 4   and   3 ◯ 3   so   3 + 3 ◯ 4 + 3

3.
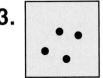

4 ◯ 3   and   3 ◯ 3   so   4 − 3 ◯ 3 − 3

4.

3 ◯ 5   and   2 ◯ 2   so   3 + 2 ◯ 5 + 2

5. Make your own.

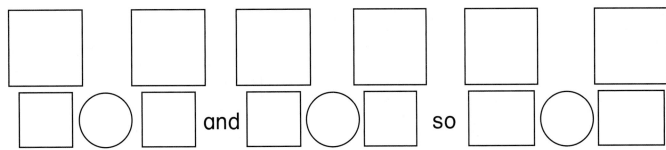

and   so

NOTE: Your child is learning what happens when the same
number is either added to or subtracted from both sides of
a number sentence.

What are the missing numbers and symbols?

6.

$16 > 15$ and $6 \bigcirc \square$

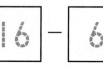

so $16 - 6 \bigcirc \square - \square$

$\square \bigcirc \square$

7.

$\square \bigcirc \square$ and $\square \bigcirc \square$

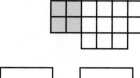

so $\square + \square \bigcirc \square + \square$

$\square \bigcirc \square$

Name _____   Date _____

Comparing and Ordering Weights

NCTM Standards 1, 2, 4, 6, 7, 8, 9, 10
TEKS 1.7F, 1.12A, 1.12B

Which words make the sentence true?
Write <, >, or =.

1.

The banana

is lighter than   the apples.

is heavier than

weighs the same as

banana ⟨ < ⟩ apples

2.

The doll   is lighter than   the teddy bear.

is heavier than

weighs the same as

doll ◯ bear

3.

The teddy bear   is lighter than   the banana.

is heavier than

weighs the same as

bear ◯ banana

NOTE: Your child is learning to use a pan balance to compare the weights of objects. Ask your child how he or she knew which object was heavier.

Write <, >, or =.

4.

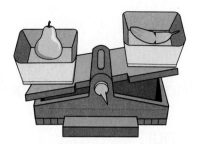

pear  feather

5.

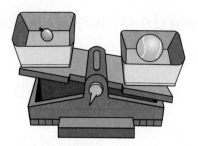

grape ◯ tennis ball

6.

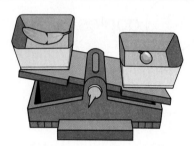

feather ◯ grape

7.

pear ◯ tennis ball

8. pear ◯ grape

9. feather ◯ tennis ball

10. Circle the lightest.

11. Circle the heaviest.

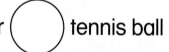

Problem Solving

12. Mari has three boxes.
Box A is heavier than Box B.
Box C is lighter than Box B.
List the boxes in order from lightest to heaviest.

_____

Changing Both Pans of a Balance

NCTM Standards 1, 2, 4, 6, 7, 8, 9, 10

TEKS 1.7F, 1.12A

What are the missing symbols and numbers?

1.

   and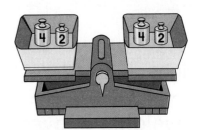

8 oz $\left(>\right)$ 4 oz   6 oz $\left(=\right)$ 6 oz

so

8 oz + 6 oz $\bigcirc$ 4 oz + 6 oz

_____ oz $\bigcirc$ _____ oz

2.

   and

_____ oz $\bigcirc$ _____ oz   _____ oz $\bigcirc$ _____ oz

so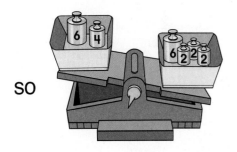

_____ oz + _____ oz $\bigcirc$ _____ oz + _____ oz

_____ oz $\bigcirc$ _____ oz

NOTE: Your child is exploring what happens when the same weight is added to both pans of a balance scale. Your child uses logic to determine which pan is heavier.

How much does each bag weigh?

3.

A weighs __10__ oz.

4.

B weighs _____ oz.

5.

C weighs _____ oz.

Use the weights above.
What are the missing numbers and symbols?

6.

_____ oz ◯ _____ oz

7.

_____ oz ◯ _____ oz

8. Circle the heaviest bag.

9. Circle the lightest bag.

Challenge
Look at the balances at the right.

10. Circle the heaviest bag.

11. Circle the lightest bag.

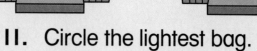

Name _____ Date _____

Problem Solving Strategy
Guess and Check

NCTM Standards 1, 2, 4, 6, 7, 8, 9, 10
 TEKS 1.1A, 1.4, 1.7F, 1.7G, 1.11B, 1.11C, 1.12A, 1.13

Understand
Plan
Solve
Check

1. Tanya has 10 fruits in a basket.
She has apples, bananas, and cherries.
The sentences at the right show how
they compare.
How many of each fruit could Tanya have?

> apples > bananas
> cherries > bananas
> cherries < apples

_____ apples   _____ bananas   _____ cherries

2. The temperature was 83°F on Monday.
It was 4°F cooler on Tuesday than on Monday.
It was warmer on Wednesday than Tuesday.
What could the temperature have been on
Wednesday?

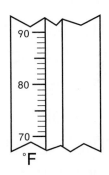

_____ °F

3. Theo used pan balances to compare how much his
book, cap, and ball weigh.

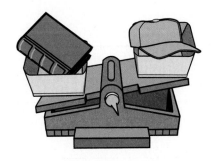  

List the objects in order from lightest to heaviest.

_____ , _____ , _____

🏠 **NOTE:** Your child is using the strategy, *guess and check,*
to solve problems. Ask your child to explain how he or she
solved the problems on this page.

Problem Solving Test Prep

1. Scott left the library at 5:30.
   He spent 2 hours there.
   What time did Scott get to
   the library?

   (A) 2:30   (C) 7:00

   (B) 3:30   (D) 7:30

2. Tina made a pattern.

   What are the next two figures
   in her pattern?

   (A) ●▲   (C) ●■

   (B) ●●   (D) ▲■

Show What You Know

3. Madison lines up 4 blocks.
   - The red block is left of the
     green block.
   - The yellow block is right of the
     blue block.
   - The green block is left of the
     blue block.

   List the blocks in order from left
   to right.

   _____

   Explain how you know your
   answer is correct.

   _____

   _____

   _____

4. Jenny has 4 cards to decorate.
   She wants to put 4 stickers on
   each card.
   She has 14 stickers.
   Does Jenny have enough
   stickers?

   _____

   Use words, numbers, or
   pictures to explain how
   you know.

   _____

   _____

   _____

   _____

   _____

   _____

Chapter 11 ## Review/Assessment

NCTM Standards 1, 2, 4, 6, 7, 8, 9, 10

1. Write < or >. Lesson 2

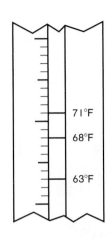

68°F ◯ 71°F

63°F ◯ 68°F

71°F ◯ 63°F

Write <, >, or =. Lessons 1 and 3

2. 9 ◯ 12	**3.** 18 ◯ 22	**4.** 37 ◯ 37
5. 52 ◯ 39	**6.** 46 ◯ 39	**7.** 15 ◯ 51

8. Write the numbers.
Then order the numbers. Lesson 4

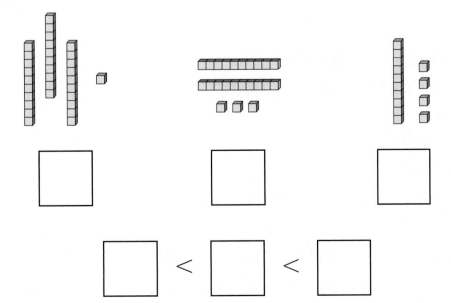

☐   ☐   ☐

☐ < ☐ < ☐

9. **Write <, >, or =.** Lesson 5

4 ◯ 6   and   2 ◯ 2   so   4 + 2 ◯ 6 + 2

How much does each bag weigh? Lessons 6 and 7

10.

A weighs _____ oz.

11.

B weighs _____ oz.

12.

C weighs _____ oz.

---

13. Circle the heaviest bag.

14. Circle the lightest bag.

Problem Solving Lesson 8

15. Dan has 7 pets.
He has dogs, cats, and fish.
The sentences tell how they compare.

dogs < cats
cats > fish
fish = dogs

How many of each pet could he have?

_____ dogs   _____ cats   _____ fish

Chapter 12

Length, Area, and Capacity
How Many Steps? ✏

You need
• index card

Walk in a straight line from the front of the room to the back.

STEP 1 Counting your steps

How many steps did you take? _____ steps

Did you take big steps or small steps? _____

Write the number of steps on a card. Then switch cards with a partner. How many steps did your partner take? _____ steps

STEP 2 Comparing steps

Compare your card with your partner's card.

Did you take the same number of steps? _____

Who took more steps? _____

Do you think your partner took big steps or small steps? Why?

_____

_____

STEP 3 Thinking About It

Why could the number of steps be different?

_____

_____

Investigation

Dear Family,

Today we started Chapter 12 in *Think Math!* In this chapter, I will measure length with nonstandard units such as paper clips, and standard units such as inches and centimeters. I will also find the area of different figures by counting the number of square units a figure covers, and compare the capacity of various containers. There are NOTES on the Lesson Activity Book pages to explain what I am learning every day.

Here are some activities for us to do together at home. These activities will help me understand length and capacity.

Love,

_____

Family Fun

Find That Length!

Play the game *Find that Length!* with your child.

- Use index cards or slips of paper to create 12 Length Cards. Each card should include a length of 1 to 12 inches.

5 inches

- Mix the cards and place them face down in a pile.

- The first player picks a card and tries to find an item in the home that is about that length. The player uses an inch ruler to measure the item to the nearest whole inch.

- If the length is about the same as that on the card, the player scores 5 points. If not, the player finds the difference between the length of the object and the length on the card. The other player gets that many points.

- The card is returned to the bottom of the deck.

- Players take turns. The first player to score 25 points wins!

Capacity in the Kitchen

Work with your child to compare the capacities of common containers.

- Show your child three clean, empty containers of various sizes and shapes. Tell your child that you will work together to see how much each container holds.

- Label the containers A, B, and C by writing each letter on a self-stick note.

- Using a paper cup as a measuring tool, find the capacity of each container. Fill the cup with water, dried beans, or rice to determine how many cupfuls each container holds.

- Count the number of cups aloud as you and your child fill each container. Have your child write down the total number of cups it takes to fill each container.

Measuring Length with Nonstandard Units

NCTM Standards 1, 2, 4, 6, 7, 8, 9, 10

TEKS 1.7A, 1.7C, 1.12A

About how long is each object? Use paper clips to measure.

Make sure all the clips are the same size

1.

about _____ paper clips

2.

about _____ paper clips

3.

Crayon

about _____ paper clips

---

4. Make your own.

about _____ paper clips

NOTE: Your child is learning to measure items using small objects, such as paper clips. Have your child measure items around the house in a similar way.

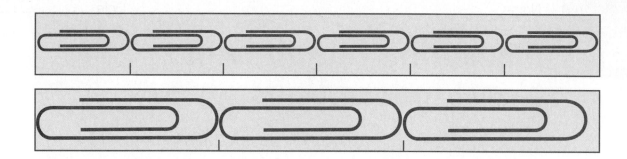

How long is each line below?
Record the length in the table.

Use paper-clip strips like these from Activity Master 64.

5. A ——————————

6. B ——————————————

7. C ————

8. D ——————————————————

Line	A	B	C	D
Small clip	about 2			
Large clip				

Problem Solving

9. What if you used larger paper clips?
How would the measurements change? Explain.

_____

_____

_____

Comparing and Ordering Lengths

NCTM Standards 1, 2, 4, 6, 7, 8, 9, 10

TEKS 1.7B, 1.12A, 1.12B

Write >, <, or =.

1.

red $\bigcirc$< purple

red + brown $\bigcirc$< purple + brown

2.

yellow $\bigcirc$ black

yellow + light green $\bigcirc$ black + light green

3.

red $\bigcirc$ dark green

red + orange $\bigcirc$ dark green + orange

4.

blue $\bigcirc$ blue

blue + purple $\bigcirc$ blue + purple

NOTE: Your child is learning to measure and compare the lengths of classroom objects, such as the rods on this page.

Write >, <, or =.

5.

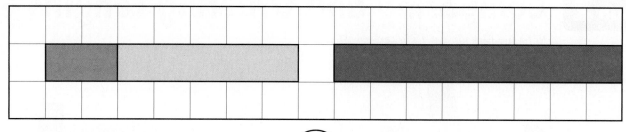

$$2 + 5 \bigcirc 8$$

6.

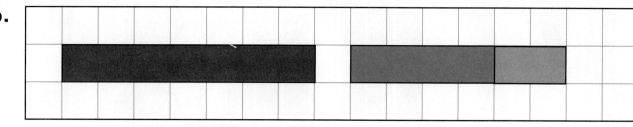

$$7 \bigcirc 4 + 2$$

7. Make your own.

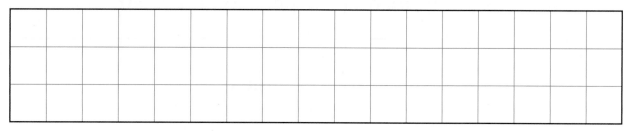

Problem Solving
8. Find these objects in your classroom.
List them in order from shortest to longest.

_____

_____

_____

Eraser

Chapter 12
Lesson 3

Measuring with a Centimeter Ruler

NCTM Standards 1, 4, 6, 7, 9, 10

Measure to the nearest centimeter.

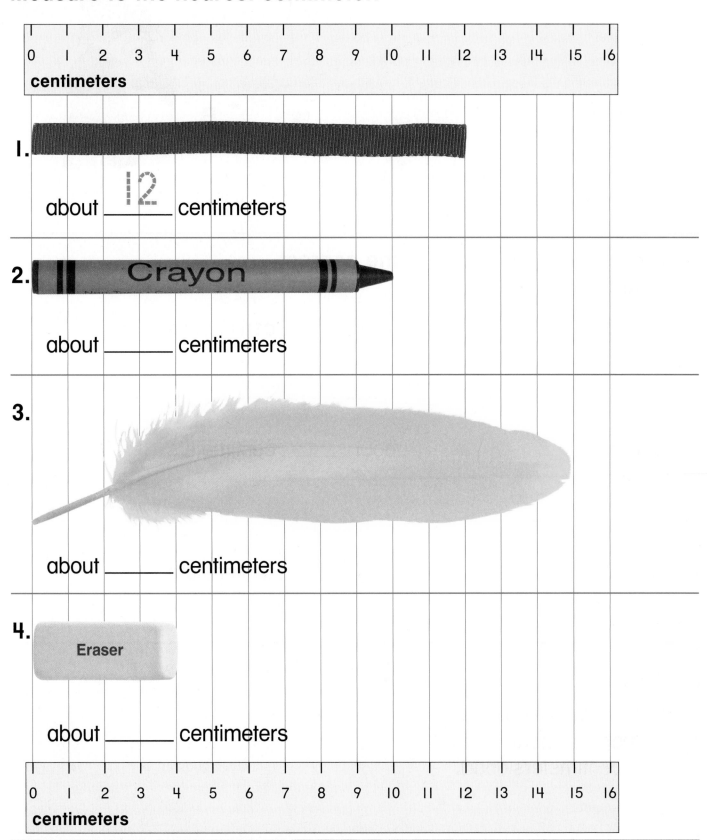

0 1 2 3 4 5 6 7 8 9 10 11 12 13 14 15 16
centimeters

1.

about __12__ centimeters

2. Crayon

about _____ centimeters

3.

about _____ centimeters

4. Eraser

about _____ centimeters

0 1 2 3 4 5 6 7 8 9 10 11 12 13 14 15 16
centimeters

© Education Development Center, Inc.

NOTE: Your child is learning to measure the length of classroom objects using a centimeter ruler. You may wish to ask your child to measure an object at home to the nearest centimeter.

5. Connect the dots.

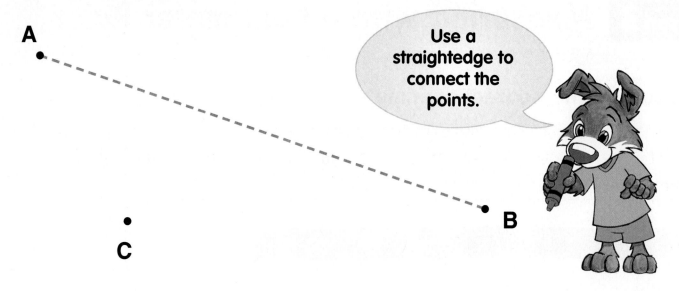

A

Use a straightedge to connect the points.

B

C

Measure each line above to the nearest centimeter.

6. A to **B**   about _____ centimeters

---

7. C to **A**   about _____ centimeters

---

8. B to **C**   about _____ centimeters

Challenge

9. Connect the dots that are about 6 centimeters apart.

Name _____ Date _____

Measuring with an Inch Ruler

NCTM Standards 1, 4, 6, 7, 8, 9, 10
TEKS 1.7B, 1.12A

Measure to the nearest inch.

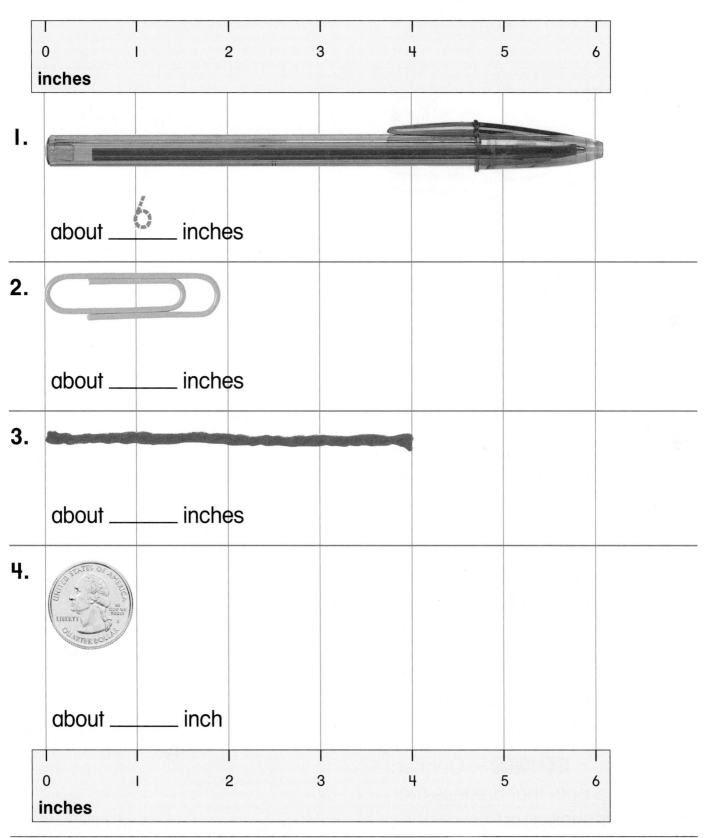

0   1   2   3   4   5   6
inches

1.

about ____6____ inches

2.

about _____ inches

3.

about _____ inches

4.

about _____ inch

0   1   2   3   4   5   6
inches

NOTE: Your child is learning to measure objects using an inch ruler. You may wish to have your child measure an object at home to the nearest inch.

Measure to the nearest inch.

5. your notebook

about _____ inches

6. your shoe

about _____ inches

7. Which is longer—your notebook or your shoe?
About how much longer?

_____

_____

_____

Challenge

8. Use a ( ) RED). Connect
two dots that are about
2 inches apart.

A.

.C

9. Use a ( ) BLUE). Connect
two dots that are more than
2 inches apart.

10. Use a ( ) GREEN). Connect
two dots that are less than
2 inches apart.

.B

Chapter 12
Lesson 5

Comparing Figures by Size

NCTM Standards 1, 2, 3, 4, 6, 7, 8, 9, 10
TEKS 1.7D, 1.12A

Connect the dots.
Measure to the nearest
centimeter.

We can use cm as a short way to write centimeter.

1. A•

C•   •B

Line	Length
A to **B**	about 5 cm
B to **C**	about _____ cm
A to **C**	about _____ cm

2. K•

M•   •L

Line	Length
K to **L**	about _____ cm
L to **M**	about _____ cm
K to **M**	about _____ cm

© Education Development Center, Inc.

NOTE: Your child is learning to measure and compare the sides of figures. Ask your child to describe the relationship between the two triangles on this page.

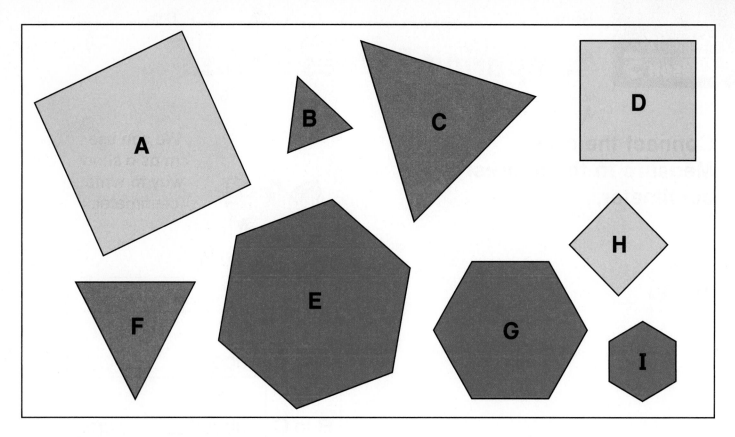

3. Complete the table.

	Largest	In-Between	Smallest
Square ◻	A		
Triangle ▲		F	
Hexagon ⬡			

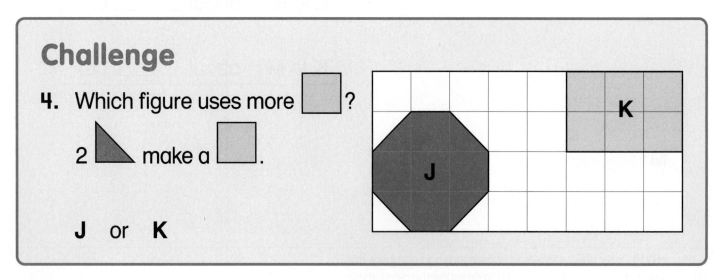

Challenge

4. Which figure uses more ◻ ?

2 ◣ make a ◻.

J or **K**

Name _____   Date _____

Exploring Area

NCTM Standards 1, 3, 4, 6, 7, 8, 9, 10

1. Connect the dots. Measure to the nearest inch.

Line	Length
A to **B**	about _____ 6 _____ inches
B to **C**	about _____ inches
A to **C**	about _____ inches

Line	Length
D to **E**	about _____ inches
E to **F**	about _____ inches
D to **F**	about _____ inches

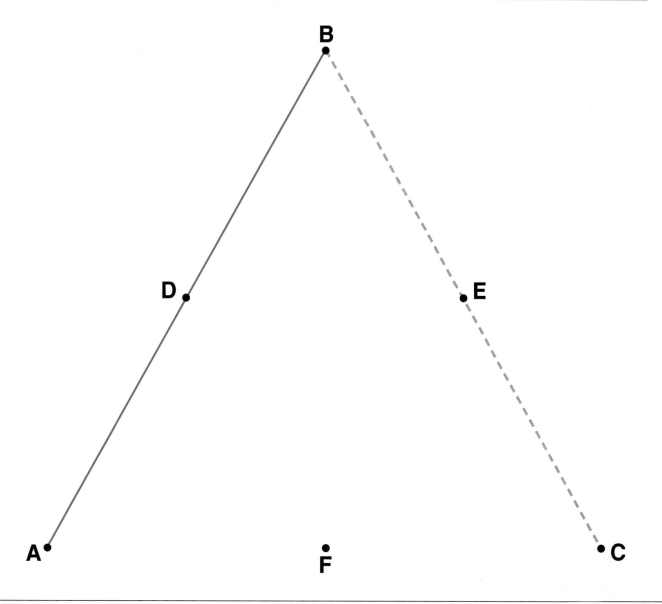

NOTE: Your child is learning to measure the sides of different figures and is exploring the concept of area. You may wish to give your child two pieces of paper in different sizes and ask, "Which piece covers more area on the table?"

2. Measure to the nearest inch.

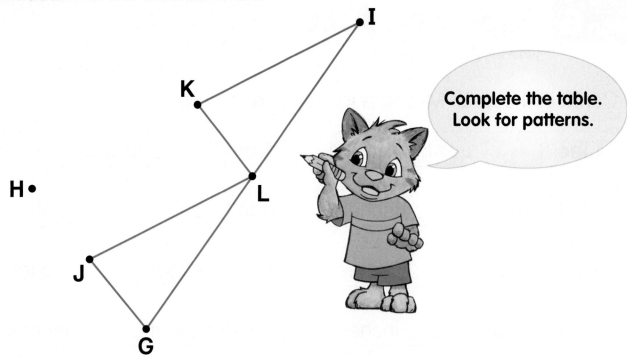

Complete the table. Look for patterns.

Line	Length
H to **G**	about ___2___ inches
H to **I**	about _____ inches
G to **I**	about _____ inches

Line	Length
H to **J**	about _____ inches
H to **K**	about _____ inches
J to **K**	about _____ inches

Problem Solving

3. Make a triangle this way:

① Draw one side.

② Draw another side twice as long.

③ Draw the third side.

Write the length of each side.

Name _____ Date _____

Finding Area on a Grid

NCTM Standards 1, 3, 4, 6, 7, 9, 10

What is the area of each shaded figure?

1.

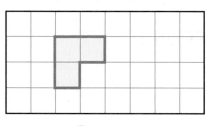

_3_ ▢

2.

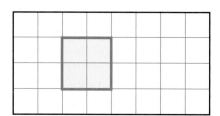

_____ ▢

3.

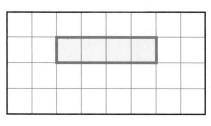

_____ ▢

4.

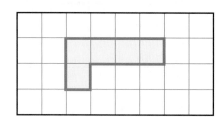

_____ ▢

5.

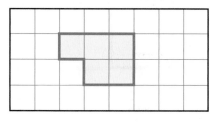

_____ ▢

6.

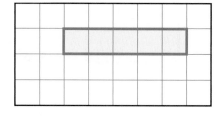

_____ ▢

7.

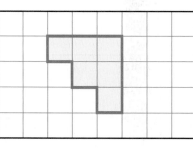

_____ ▢

8.

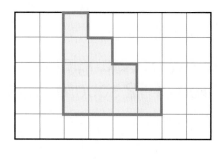

_____ ▢

NOTE: Your child is learning to find the area of figures by counting square tiles or squares on a grid. You may wish to ask your child to draw two different figures with the same area on grid paper.

49 CCXLV two hundred forty-five 245

What is the area of each shaded figure?

9.

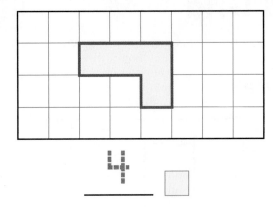

__4__ ☐

10.

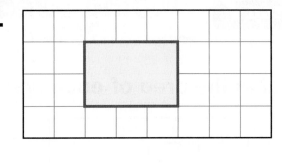

_____ ☐

11.

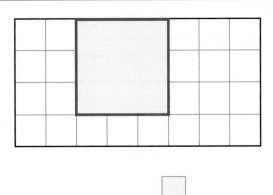

_____ ☐

12.

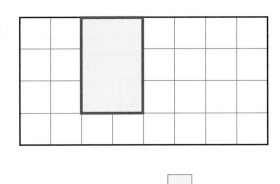

_____ ☐

13.

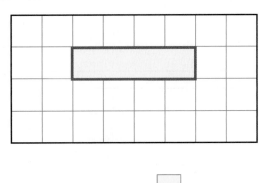

_____ ☐

14.

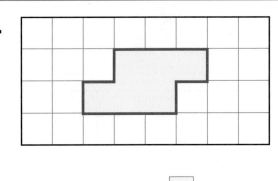

_____ ☐

Challenge

15. Draw a rectangle with an area of 12 square units.

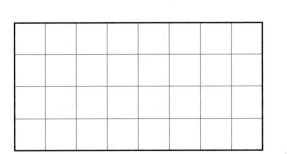

Chapter 12
Lesson 8 **Comparing Areas**

NCTM Standards 1, 3, 4, 6, 7, 9, 10

🔶 **TEKS 1.7D**

What is the area of each figure?

1.
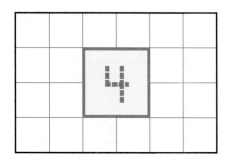

2.
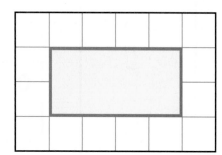

3.
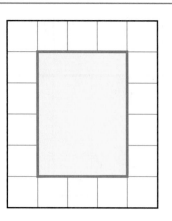

4.
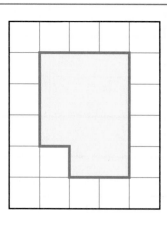

5.
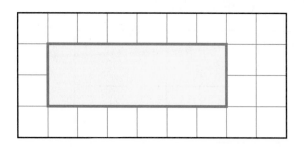

6.
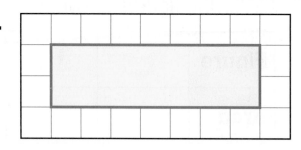

7.
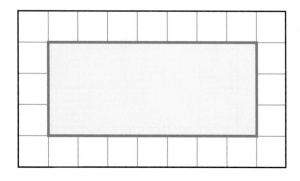

8.
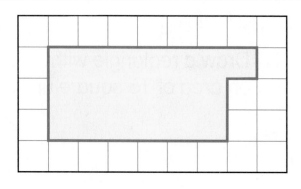

NOTE: Your child is learning to find the area of figures using
the area of smaller figures, and to compare areas.

100 + 140 + 7 **CCXLVII** two hundred forty-seven **247**

9. Write the area on each figure.

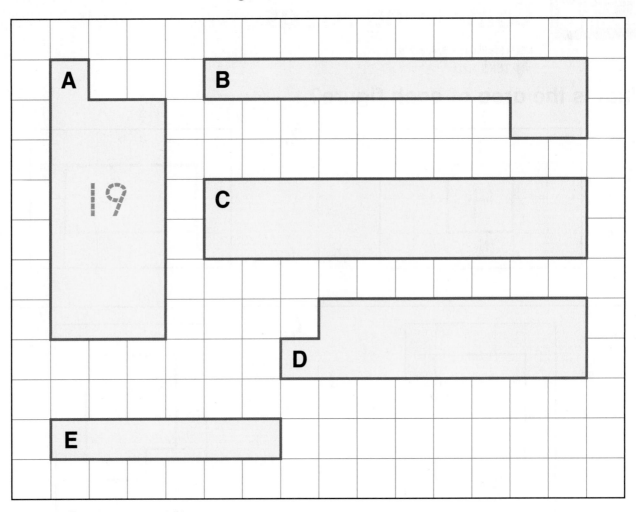

---

10. Order the figures above from least to greatest area.

Figure	E	B			
Area	6				

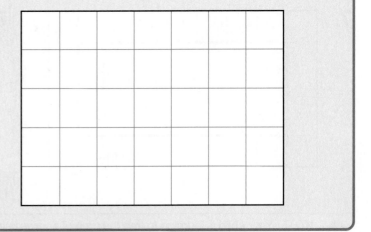

Challenge

11. Draw a rectangle with an area of 15 square units.

Measuring Boxes and Rectangles

NCTM Standards 1, 3, 4, 6, 7, 8, 9, 10

TEKS 1.12A

What is the length of each side?

Remember, cm is short for centimeter.

1.

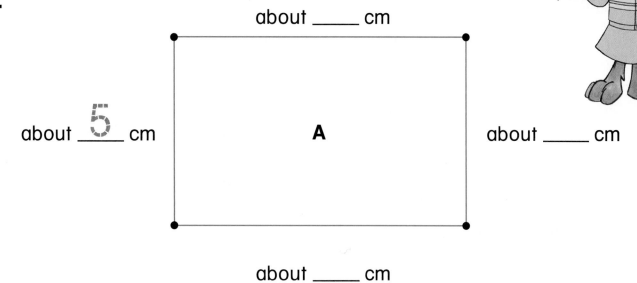

about _____ cm

about 5 cm

A

about _____ cm

about _____ cm

2.

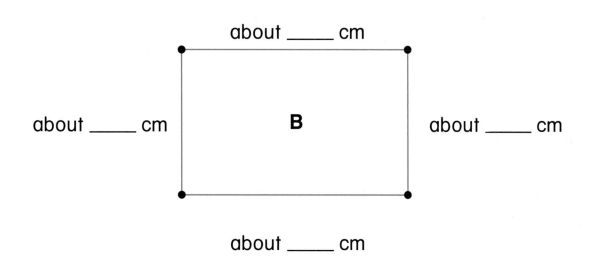

about _____ cm

about _____ cm

B

about _____ cm

about _____ cm

3. Will Rectangle B fit inside Rectangle A? Explain.

_____

_____

NOTE: Your child is learning to measure and compare measurements. You might ask your child to measure the length, width, and height of two boxes and decide whether one will fit inside the other.

What is the length of each side?

4.

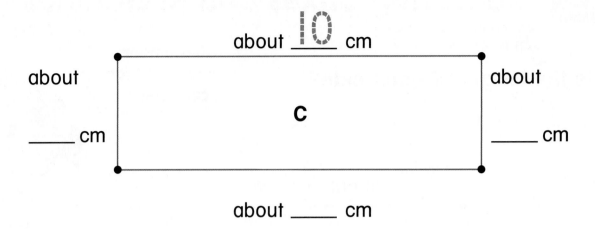

about __10__ cm

about _____ cm

C

about _____ cm

about _____ cm

5.

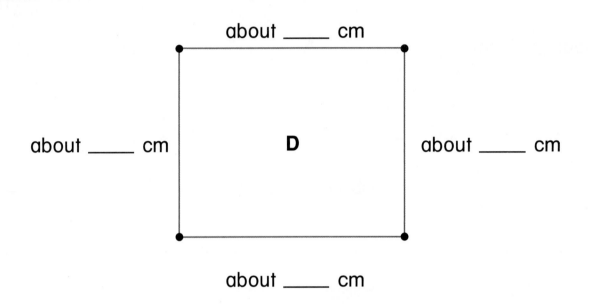

about _____ cm

about _____ cm

D

about _____ cm

about _____ cm

Challenge

6. Will Rectangle C fit inside Rectangle D? _____

Will Rectangle D fit inside Rectangle C? _____
Explain.

_____

_____

_____

Introducing Capacity with Nonstandard Units

NCTM Standards 1, 4, 6, 7, 8, 9, 10

🔶 TEKS 1.7E, 1.10A, 1.12A

Carol uses scoops to measure how much each container will hold.

The bar graph shows how many scoops fit.

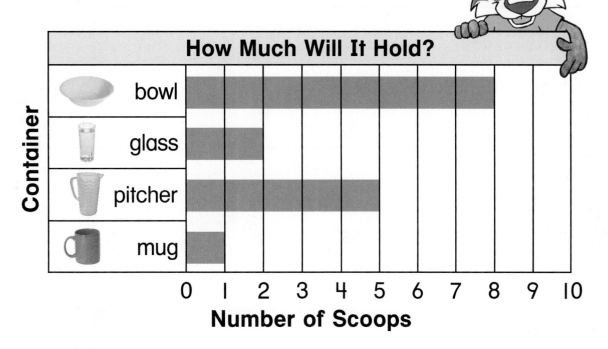

How Much Will It Hold?

Container — bowl, glass, pitcher, mug

Number of Scoops — 0 1 2 3 4 5 6 7 8 9 10

1. Which container holds the most? ___bowl___

2. Which container holds the least? _____

✏️ **3.** How many more scoops does the bowl hold than the pitcher? Explain.

_____

_____

_____

NOTE: Your child is exploring capacity by filling containers. You might have your child fill two containers with water to decide which holds more.

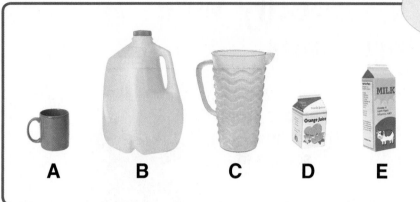

Compare how much they hold.

A   B   C   D   E   F

---

4. Which do you think holds more?   or

---

5. Which do you think holds less?      or

---

6. Look at **A**, **B**, and **C**.
List them in order from holds
the least to holds the most.

_____ _____ _____
least   most

---

7. Look at **C**, **D**, and **E**.
List them in order from holds
the most to holds the least.

_____ _____ _____
most   least

Problem Solving

8. José has 3 jars labeled X, Y, and Z.
Use the clues.
Which container holds the least?

_____

Clues
X holds more than Z.
Y holds less than X.
Z holds more than Y.

Measuring Capacity with Standard Units

NCTM Standards 1, 4, 6, 7, 8, 9, 10

 TEKS 1.13

Which is the better measurement?

I.

(I quart)   10 quarts

2.

6 pints   60 pints

3.

2 gallons   20 gallons

4.

I liter   10 liters

 5. Draw your own container.
Tell about it. How much will it hold?

NOTE: Your child is learning to compare the capacities of different containers and to identify appropriate measurements and measuring tools.

What could the real measurement be?

6.

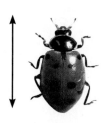

(2 centimeters)   2 gallons

7.

8 inches   8 pints

8.

less than   more than
I pint   I gallon

9.

less than   more than
I pint   I gallon

Challenge

10. There are 2 pints in I quart.
There are 4 quarts in I gallon.
How many pints are in I gallon?

_____ pints

Use words,
numbers, or pictures
to explain.

Problem Solving Strategy
Draw a Picture

NCTM Standards 1, 4, 6, 7, 8, 9, 10

TEKS 1.7E, 1.11A, 1.11B, 1.11C

Understand
Plan
Solve
Check

I. A frog is at the bottom of a hole. The hole is 10 inches deep. The frog jumps up 3 inches each time. How many jumps will it take to get to the top?

_____ jumps

2. Scott has 6 square tiles. How many different rectangles can he make with the tiles?

_____ rectangles

3. Heidi has a bowl, a drinking cup, and a pitcher.
The pitcher holds more than the bowl.
The bowl holds 2 cups.
List the containers in order from holds the most to holds the least.

_____

NOTE: Your child is exploring different ways to solve problems. Drawing a picture can help children visualize patterns and relationships in a problem.

Problem Solving Test Prep

1. George has 2 gallons of punch. He wants to fill 4 pitchers. Each pitcher holds 1 quart. There are 4 quarts in 1 gallon. How much punch will he have left?

(A) 1 gallon   (C) 2 quarts

(B) 3 quarts   (D) 5 quarts

2. Brooke uses square tiles to make this pattern.

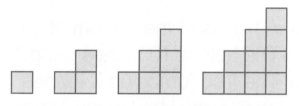

How many tiles does she need for the next figure?

(A) 11   (C) 16

(B) 15   (D) 21

 Show What You Know

3. Joey uses 5 inches of string to make a bookmark. He makes 6 bookmarks. How many inches does he use?

_____ inches

Explain.

_____

_____

_____

_____

4. Madison has some dimes and pennies. She has 47¢. How many dimes and pennies could she have?

_____ dimes

_____ pennies

Explain.

_____

_____

_____

_____

Chapter 12 # Review/Assessment
NCTM Standards 1, 3, 4, 6, 9, 10

1. About how long is the line?
Lesson 1

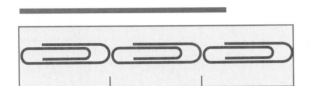

about _____ paper clips

2. Write >, <, or =. Lesson 2

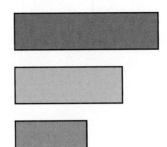

purple ◯ **green**

3. About how many centimeters long is the paper clip? Lesson 3

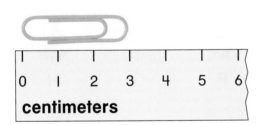

about _____ centimeters

4. About how many inches long is the eraser? Lesson 4

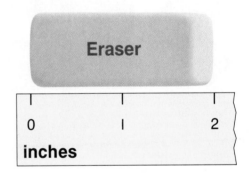

about _____ inches

5. Which figure is the smallest?
Lesson 5

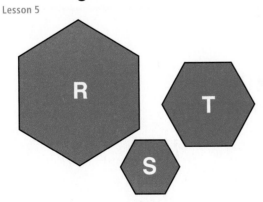

Figure _____

6. Draw a figure with a smaller area. Lesson 6

What is the area of each figure? <small>Lessons 7 and 8</small>

7.

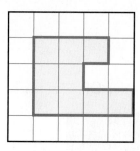

_____ ☐

8.

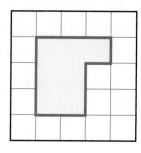

_____ ☐

9. Which figure will fit inside of the other? <small>Lesson 9</small>

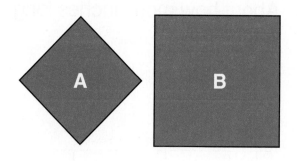

_____

10. Order the containers from holds the least to holds the most. <small>Lessons 10 and 11</small>

F   G   H

_____

Problem Solving <small>Lesson 12</small>

11. Elena makes a quilt with 12 squares patches. The patches are all the same size. How many rectangles can she make?

_____ rectangles

Name _____

Making and Breaking Numbers

How Many Ways? ✏️

STEP 1 Counting

Use cubes to make a train.
How many cubes are in your train?

_____

STEP 2 Adding

Break your train into 2 trains. Find as many different
ways as you can. Write a number sentence for each way.

_____

_____

_____

_____

How many ways did you find? _____ ways

STEP 3 Comparing

Share what you did with your classmates.
What do you notice?

_____

_____

Investigation

Dear Family,

Today we started Chapter 13 in *Think Math!* In this chapter, I will learn about even and odd numbers. I will learn how to break apart numbers to find sums to 20. There are NOTES on the Lesson Activity Book pages to explain what I am learning every day.

Here are some activities for us to do together at home. These activities will help me learn to create reasonable story problems and to make change.

Love,

_____

Family Fun

Even or Odd?

Your child will be familiar with this game from Lesson 13.1.

- You will need a copy of Activity Master 79: Even or Odd.

- Each player chooses *even* or *odd*.

- Both you and your child say a number aloud at the same time. Write your number and whether it is even or odd in the left box. Write your child's number and whether it is even or odd in the middle box. Then, have your child record whether the sum will be even or odd in the right box.

- If the sum is even, the *even* player gets a point. If the sum is odd, the *odd* player gets a point. Continue play until one of you earns 3 points.

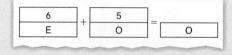

What's the Addend?

Work with your child to practice finding sums to 20.

- Gather 11 index cards or slips of paper to make a deck of sum cards. Write the numbers 10 to 20 (one number per card) on the cards. Shuffle the cards and turn them facedown in a pile.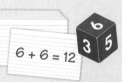

- One player chooses a sum card and reads the number aloud. The other player tosses a number cube and says the number aloud.

- Work with your child to find the number that needs to be added to the number tossed to make the sum on the card.

- Switch roles and repeat. As your child becomes more comfortable with the game, pick up the speed and ask your child to call out the missing number as quickly as possible.

Chapter 13
Lesson 1

Making Even and Odd Numbers

NCTM Standards 1, 2, 6, 7, 8, 9, 10

TEKS 1.3B, 1.5B, 1.12A

Draw rods to show the sum.
Is the sum even or odd?

Color to show
the red rods.
■ stands for 2.
□ stands for 1.

1. 2 + 3   even   (odd)

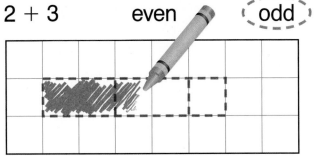

2. 5 + 5   even   odd

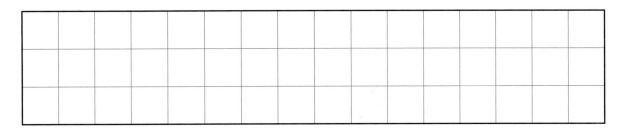

3. 6 + 8   even   odd

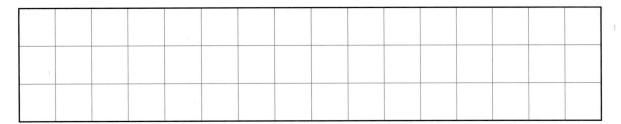

 4. Is 4 + 7 even or odd? Explain.

_____

_____

_____

 NOTE: Your child is learning to add even and odd numbers.
Ask your child to explain why the sum of 2 and 3 is an
odd number.

Draw a picture for each number sentence.
Is the sum even or odd?

5. even + even = _____

_____

6. odd + odd = _____

_____

7. even + odd = _____

Challenge
Draw a picture for each.
Is the sum even or odd?

8. odd − even = _____

9. even − odd = _____

Name _____ Date _____

Making Numbers as Sums of 1, 2, 4, and 8

NCTM Standards 1, 2, 6, 7, 8, 9, 10

TEKS 1.3A, 1.12A

What number does each rod show?

1.

_____

2.

_____

3.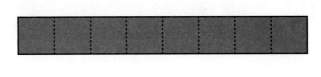

_____

4.

_____

What number sentence does each train show?

5.

$\underline{\quad 2 \quad} + \underline{\quad 4 \quad} = \underline{\qquad}$

6.

$\underline{\qquad} + \underline{\qquad} = \underline{\qquad}$

7.

$\underline{\qquad} + \underline{\qquad} = \underline{\qquad}$

8.

$\underline{\qquad} + \underline{\qquad} = \underline{\qquad}$

 NOTE: Your child is learning to combine the numbers 1, 2, 4, and 8 to make larger numbers. Ask your child to find a way to make 5 using these numbers.

What number sentence does each train show?

9.

_____ + _____ + _____ = _____

10.

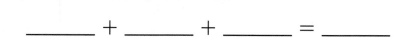

_____ + _____ + _____ = _____

11.

_____ + _____ + _____ = _____

12.

_____ + _____ + _____ + _____ = _____

 13. What other number can you make with 3 rods? Explain.

_____

_____

Challenge

Use the numbers 1, 2, 4, 8, and 16 to complete the number sentences.

14. ____ + ____ = 17   **15.** ____ + ____ = 18

16. ____ + ____ + ____ = 19   **17.** ____ + ____ = 20

Combining Triangular Numbers

NCTM Standards 1, 2, 6, 7, 8, 9, 10

TEKS 1.3B, 1.5E, 1.12A

How many dots are in each triangle?

1.

2.
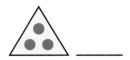 ____

3.
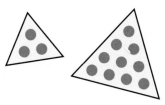 ____

4.
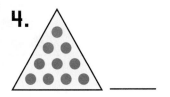 ____

What is the fact family?

5.

$3 + 10 = 13$

$10 + 3 = 13$

$13 - 3 = 10$

$13 - 10 = 3$

6.
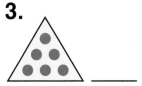

____ + ____ = ____

____ + ____ = ____

____ - ____ = ____

____ - ____ = ____

7.

____ + ____ = ____

____ + ____ = ____

____ - ____ = ____

____ - ____ = ____

8.

____ + ____ = ____

____ + ____ = ____

____ - ____ = ____

____ - ____ = ____

NOTE: Your child is learning to combine triangular numbers such as 1, 3, 6, and 10. Ask your child to use two of these numbers to make a fact family.

Write a number sentence.

9.

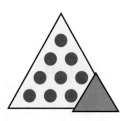

$$\boxed{10} - \boxed{1} = \boxed{9}$$

10.

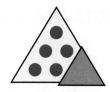

$$\boxed{} - \boxed{} = \boxed{}$$

11.

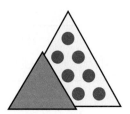

$$\boxed{} - \boxed{} = \boxed{}$$

12.

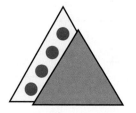

$$\boxed{} - \boxed{} = \boxed{}$$

Challenge

13. What is the next triangular number? 1, 3, 6, 10, 15, 21, _____

14. Complete the number sentences with triangular numbers. Use + or −.

16 = _____

16 = _____

Name _____ Date _____

Making Sums of 60

NCTM Standards 1, 2, 6, 7, 8, 9, 10

🔺 TEKS 1.5A, 1.12A

Only use the numbers 0, 10, 20, 30, 40, 50, and 60.

Complete each Cross Number Puzzle. Use multiples of 10.

1.

30		
		60

2.

	0	
		60

3.

20		
		60

4.

	10	
		60

NOTE: Your child is learning to add multiples of ten. You may wish to ask your child to find a different way to solve Problem 4.

5. Continue the pattern. Use these numbers to complete the Cross Number Puzzle.

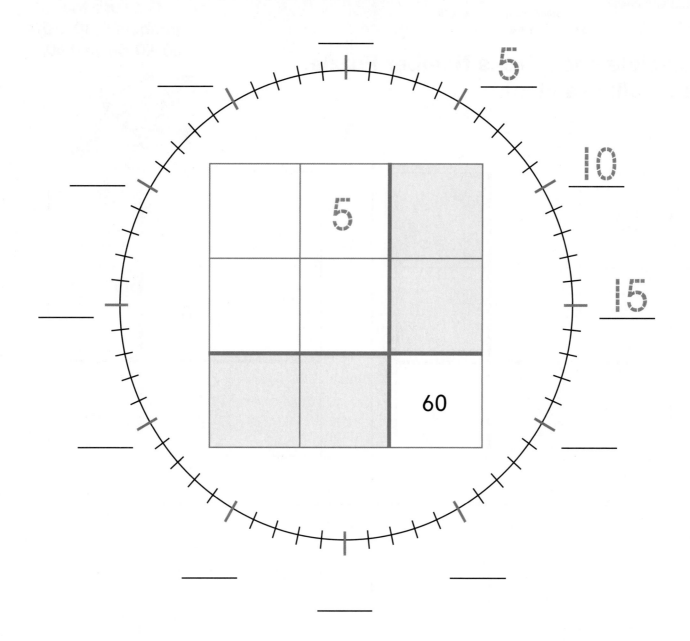

Challenge

6. Use multiples of 5 or 10 to complete this Cross Number Puzzle.

Name _____   Date _____

Sums to 12

NCTM Standards 1, 2, 6, 7, 8, 9, 10

TEKS 1.3A, 1.3B, 1.12A

Draw rods to show each sum.
What is the sum?

1. 5 + 7 = **12**

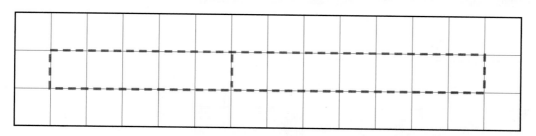

2. 8 + 3 = _____

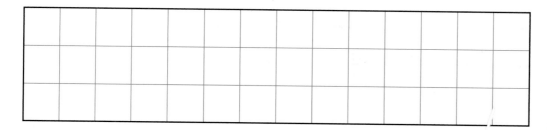

3. 4 + 8 = _____

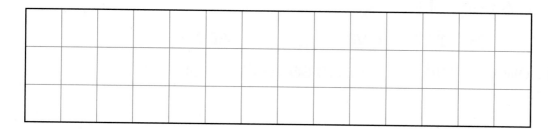

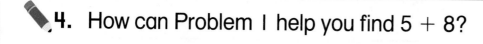**4.** How can Problem 1 help you find 5 + 8?

_____

_____

NOTE: Your child is learning to find pairs of numbers with a sum of 12. You may ask your child to list a few pairs of numbers with a sum of 12.

What number sentence does each train show?

5.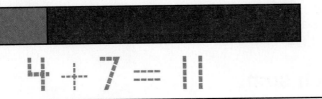

$4 + 7 = 11$

6.

_____

7.

_____

8.

_____

Problem Solving

9. Dante had 6 pairs of socks.

How many socks did he have?   _____ socks

Explain how you can use rods to solve this problem.

_____

_____

_____

Sums to 15

NCTM Standards 1, 2, 6, 7, 8, 9, 10

TEKS 1.3A, 1.3B, 1.12A

What number sentence does each train show?

1.

$$4 + 9 = 13$$

2.

_____

3.

_____

4.

_____

5. Draw rods to show 13, 14, or 15.

Write a number sentence.

_____

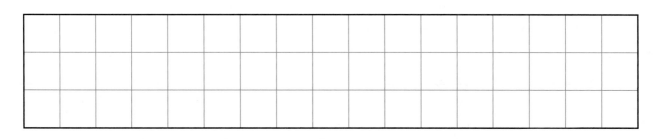

NOTE: Your child is learing to break apart numbers to make them easier to add. Ask your child to explain how to find 8 + 6 by making a group of ten.

Draw dots to show each addend.
Write the missing numbers.

6. $8 + 7$

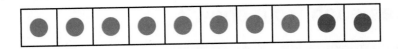

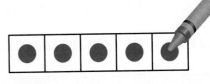

10   +   _____ = _____

7. $5 + 9$

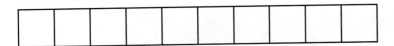

10   +   _____ = _____

8. $7 + 6$

10   +   _____ = _____

Problem Solving

9. Jenna has 8 toy animals in a case on her wall. She has 6 more on her bed. How many animals does she have? Explain how you know.

_____ animals

_____

_____

_____

Name _____ Date _____

Sums to 16

NCTM Standards 1, 2, 6, 7, 8, 9, 10
TEKS 1.3B, 1.4, 1.12A

Write the missing numbers.

1.

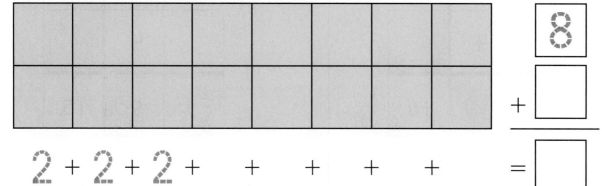

$2 + 2 + 2 + \quad + \quad + \quad + \quad + \quad =$ ☐

8

$+$ ☐

2.

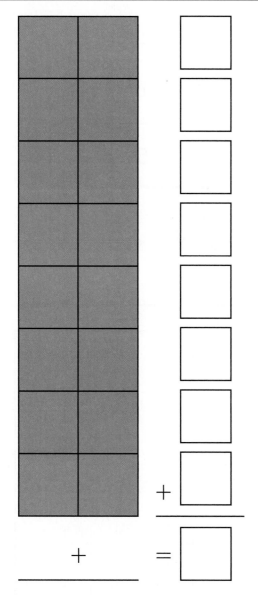

$+$ ☐

$+$ $=$ ☐

3.

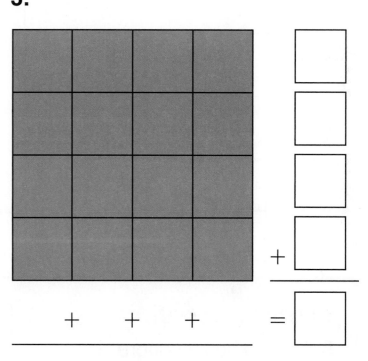

$+ \quad + \quad +$

$+$ ☐

$=$ ☐

NOTE: Your child is finding different ways to make 16.
You may wish to ask your child to use pennies to show
different ways.

Complete each Cross Number Puzzle.

4.

		5
	4	
12		16

5.

3		
	4	
	9	15

6.

	3	8
2		
	9	

7.

	4	
	5	7
5		

Problem Solving

8. Karyn has 8 square tiles. She wants to make a tray. How many different rectangles can she make?

_____

Draw them on the grid.

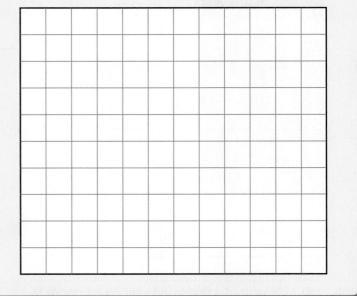

Name _____ Date _____

Sums to 18
NCTM Standards 1, 2, 6, 7, 8, 9, 10
🤚 TEKS 1.3B, 1.4, 1.12A

Write the missing numbers.

1.

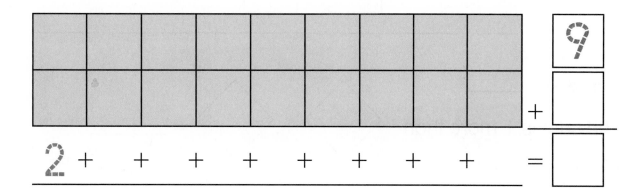

$2 + + + + + + + + = \square$

$9 + \square = \square$

2.

$+ + + + + = \square$

$\square + \square = \square$

© Education Development Center, Inc.

NOTE: Your child is learning to make sums of 18 and to use doubles facts to find near doubles. Ask your child what doubles fact can be used to find 9 + 8.

What doubles facts can you use to find the sum?

3. $6 + 7 = $ _____

The sum is 1 more than 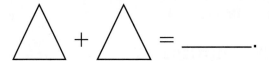 $\triangle$ + $\triangle$ = _____.

The sum is 1 less than $\square$ + $\square$ = _____.

---

4. $7 + 8 = $ _____

The sum is 1 more than 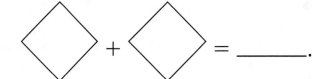 $\diamondsuit$ + $\diamondsuit$ = _____.

The sum is 1 less than  + = _____.

---

5. How would you use a doubles fact to find $9 + 8$?

_____

_____

_____

Challenge

6. How many rectangles can you draw that show 18?

Use words, numbers, or pictures to explain your answer.

_____ rectangles

Name _____ Date _____

Sums to 20

NCTM Standards 1, 2, 6, 7, 8, 9, 10

TEKS 1.3B, 1.5D, 1.12A

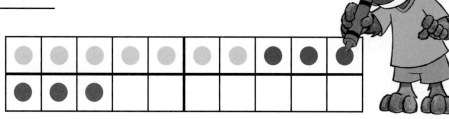

You may use counters to help.

Draw dots to show each number. Then find the sum.

1. $7 + 6 =$ ___13___

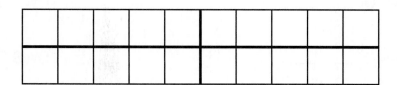

2. $8 + 7 =$ _____

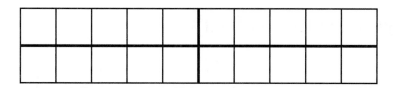

3. $9 + 9 =$ _____

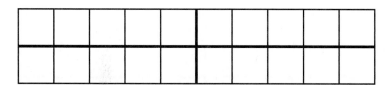

4. $10 + 10 =$ _____

NOTE: Your child is continuing to make groups of five and ten to make adding easier. Ask your child to explain how he or she solved Problem 2.

What is the sum?

5. $6 + 8 =$ _____

6. $5 + 9 =$ _____

7. $9 + 10 =$ _____

8. $7 + 5 =$ _____

9. $6 + 6 =$ _____

10. $8 + 8 =$ _____

11. $4 + 9 =$ _____

12. $7 + 7 =$ _____

13. $6 + 9 =$ _____

 14. Explain how you solved Problem 13.

_____

_____

_____

Challenge

15. What rectangles can you make with 20 tiles?

_____ rows of _____

_____ rows of _____

_____ rows of _____

_____ rows of _____

_____ rows of _____

_____ rows of _____

Problem Solving Strategy

Solve a Simpler Problem

NCTM Standards 1, 2, 4, 6, 7, 8, 9, 10

🔽 **TEKS 1.5A, 1.11B, 1.11C, 1.12A, 1.12B, 1.13**

I. Kendall traced jumps on a number line.
She jumped in a pattern.

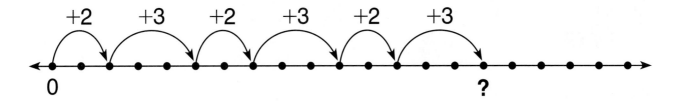

+2   +3   +2   +3   +2   +3

0   ?

If she started at 0, where did she land?

_____

---

2. Jack baked 2 batches of muffins.
He baked 6 muffins in a small tin and
9 muffins in a large tin for each batch.

How many muffins did Jack bake in all?

_____ muffins

---

3. Billie used 10 tiles to make this rectangle.

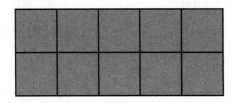

She wants to make a rectangle with 20 tiles.
How many rows and columns could her
rectangle have?

_____ rows and _____ columns

NOTE: Your child is using the strategy, *solve a simpler problem*, to solve problems. Ask your child to explain how he or she solved the problems on this page.

Problem Solving Test Prep

1. Mary made a design with tiles.

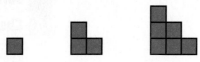
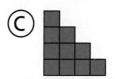
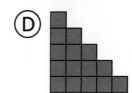

Which figure is next?

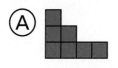
Ⓐ

Ⓒ

Ⓑ

Ⓓ

2. Josh had 12 marbles.
He lost some of them.
Now he has 6 marbles left.
How many marbles did
Josh lose?

Ⓐ 6

Ⓑ 12

Ⓒ 18

Ⓓ 24

✏ Show What You Know

3. Kari, Rico, and Jordan
ran a race.
Jordan beat Rico.
Kari beat Jordan.
Who won the race?

_____

Explain how you know your
answer is correct.

_____

_____

_____

4. There are 20 children in
Mrs. Park's class. There are
4 more boys than girls.
How many boys and girls
are in the class?

_____ boys _____ girls

Explain how you solved the
problem.

_____

_____

_____

Chapter 13 Review/Assessment
NCTM Standards 1, 2, 6, 7, 8, 9, 10

1. Write even or odd. Lesson 1

even + even = _____

---

2. What number sentence does this train show? Lesson 2

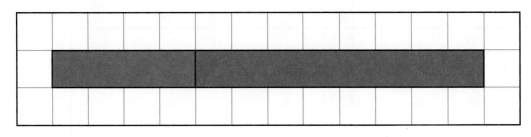

_____ + _____ = _____

What is the fact family? Lesson 3

3.

_____ + _____ = _____

_____ + _____ = _____

_____ − _____ = _____

_____ − _____ = _____

4.

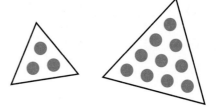

_____ + _____ = _____

_____ + _____ = _____

_____ − _____ = _____

_____ − _____ = _____

Complete each Cross Number Puzzle.
Use multiples of 10. Lesson 4

5.

30		
	10	
		60

6.

20		
	0	
		60

7. Draw rods to show a sum of 12. Lessons 5 and 6

What is the sum? Lessons 5–9

8. 9 + 8 = _____

9. 8 + 8 = _____

10. 7 + 9 = _____

Problem Solving Lesson 10

11. Nikki thought of the number pattern.

The first five numbers are 2, 4, 6, 8, 10.

What will the tenth number in her pattern be? _____

Name _____

Extending Addition and Subtraction
Adding Cents ✏️

You need
• a handful of pennies

STEP 1 Counting

How much money do you have? _____ ¢

How did you count the pennies? _____

_____

STEP 2 Adding

Add 3 nickels to the pennies.

How much money do you have now? _____ ¢

Explain how you found the total. _____

_____

STEP 3 Buying

Which of these items can you buy with your money? Explain.

 64¢

24¢

 12¢

19¢

_____

_____

_____

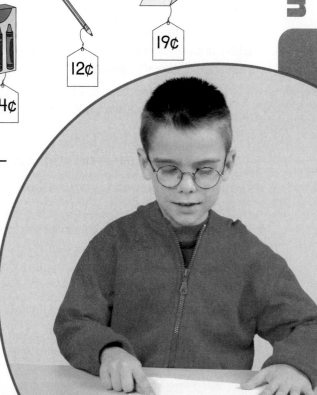

Investigation

Dear Family,

Today we started Chapter 14 in *Think Math!* In this chapter, I will learn different ways to add and subtract. I will learn how to make change from a dollar and to solve Cross Number Puzzles that have more than one solution. There are NOTES on the Lesson Activity Book pages to explain what I am learning every day.

Here are some activities for us to do together at home. These activities will help me learn to create reasonable story problems and to make change.

Love,

_____

Family Fun

Story Time

Work with your child to create story problems.

- Create two story problems. Make one problem a silly or unreasonable story and the other a reasonable story. For example,

    Jack is twice as tall as his father.
    His father is 8 feet tall. (unreasonable)

    Jack is half as tall as his father.
    His father is 6 feet tall. (reasonable)

- Discuss what makes each story reasonable or unreasonable.

- You may wish to create more stories like these and have your child make a book of math stories. Your child can draw pictures to go with each story and explain how to solve each one.

Money, Money, Money

Work with your child to practice identifying coins and making change from a dollar.

- You will need one dollar bill, pennies, nickels, and dimes. Write various amounts of money, less than a dollar, each on a separate slip of paper.

- Give your child the dollar bill. Have your child select one of the slips of paper. Ask your child to identify the amount.

- Have your child use the dollar to pretend to buy an item for the amount on the slip of paper. Work together to figure out how much change your child should get.

Name _____ Date _____

Adding Number Sentences

NCTM Standards 1, 2, 6, 7, 8, 9, 10

🔻 TEKS 1.3A, 1.11D, 1.12A, 1.13

Write the missing numbers and signs.

You may use counters or blocks to help you find sums.

1.

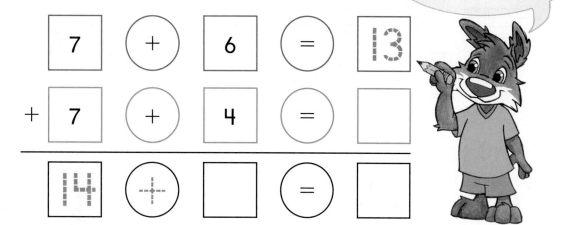

$$7 \quad + \quad 6 \quad = \quad 13$$
$$+ \quad 7 \quad + \quad 4 \quad = \quad \square$$
$$14 \quad + \quad \square \quad = \quad \square$$

2.

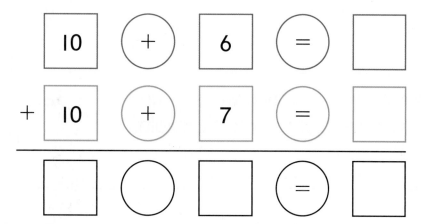

$$10 \quad + \quad 6 \quad = \quad \square$$
$$+ \quad 10 \quad + \quad 7 \quad = \quad \square$$
$$\square \quad \bigcirc \quad \square \quad = \quad \square$$

3.

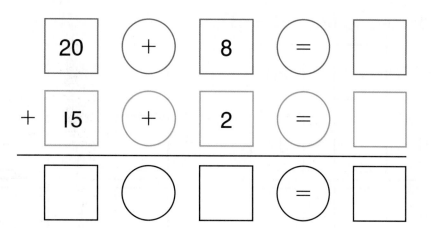

$$20 \quad + \quad 8 \quad = \quad \square$$
$$+ \quad 15 \quad + \quad 2 \quad = \quad \square$$
$$\square \quad \bigcirc \quad \square \quad = \quad \square$$

NOTE: Your child is exploring how to add two addition sentences. You may wish to ask your child to explain how he or she solved one of the problems above.

Write the missing numbers and signs.

4.

| 13 | + | 8 | = | |

| + 7 | + | 8 | = | |

| 20 | (+) | | = | |

5.

| 14 | + | 6 | = | |

| + 14 | + | 5 | = | |

| | ( ) | | = | |

6.

| 10 | + | | = | 19 |

| + | + | 2 | = | 32 |

| | ( ) | | = | |

7.

| | + | 16 | = | 19 |

| + | + | 4 | = | 32 |

| | ( ) | | = | |

 8. Problems 6 and 7 show two ways to find 19 + 32.
Which way was easier for you? Explain.

Challenge

9. Show two ways to find 28 + 37.

| | + | | = | 28 |

| + | + | | = | 37 |

| | + | | = | |

| | + | | = | 28 |

| + | + | | = | 37 |

| | + | | = | |

Name _____ Date _____

Making Addition Easier

NCTM Standards 1, 2, 6, 7, 8, 9, 10

Write the missing numbers and signs.

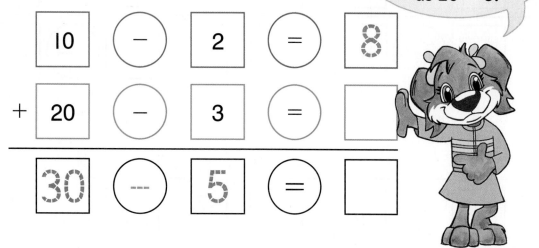

To add 8 and 17,
think of 8 as 10 − 2
and think of 17
as 20 − 3.

1.

10	−	2	=	8
+ 20	−	3	=	
30	−	5	=	

2.

20	−	1	=	
+ 20	−	4	=	
			=	

3.

30	−	2	=	
+ 20	−	4	=	
			=	

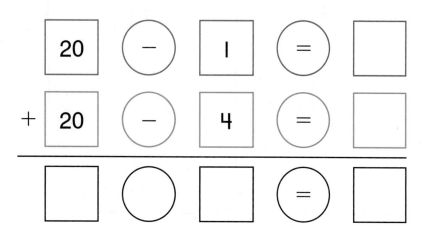

NOTE: Your child is learning to break apart numbers to make them easier to add. You might ask your child how to make 28 + 19 easier to add.

Write the missing numbers and signs.

4.

| 39 | − | 0 | = | ☐ |

| + 30 | − | 2 | = | ☐ |

$$69 \;\bigcirc\!(-)\; ☐ \;=\; ☐$$

5.

| 40 | − | 1 | = | ☐ |

| + 30 | − | 2 | = | ☐ |

$$☐ \;\bigcirc\; ☐ \;=\; ☐$$

6.

| 30 | − | 3 | = | ☐ |

| + 70 | − | 2 | = | ☐ |

$$☐ \;\bigcirc\; ☐ \;=\; ☐$$

7.

| 27 | − | ☐ | = | 27 |

| + 70 | − | 2 | = | ☐ |

$$☐ \;\bigcirc\; ☐ \;=\; ☐$$

Challenge

8. Show two ways to find 28 + 36.

| ☐ | − | ☐ | = | 28 |

| + ☐ | − | ☐ | = | 36 |

$$☐ \;\bigcirc\!(-)\; ☐ \;=\; ☐$$

| ☐ | − | ☐ | = | 28 |

| + ☐ | − | ☐ | = | 36 |

$$☐ \;\bigcirc\!(-)\; ☐ \;=\; ☐$$

Chapter 14
Lesson 3

Modeling Number Sentences and Stories

NCTM Standards 1, 2, 6, 7, 8, 9, 10

TEKS 1.3B, 1.12A

Write the number sentence and answer the question.

1. There are 7 children in the tent.

How many children are camping?

$$\boxed{7} + \boxed{3} = \boxed{}$$

_____ children

---

2. There are 23 children on each bus.

How many children are on the buses?

$$\boxed{} + \boxed{} = \boxed{}$$

_____ children

NOTE: Your child is learning how to solve problems in different ways. You might ask your child to explain how to solve problem 2.

© Education Development Center, Inc.

Read the problem. Write the number sentence and answer each question.

It took Devi 14 days to read 8 books. Then he read 5 more books in 1 week.

3. How many books did Devi read?

_____ books

4. How many days did it take?

_____ books

✏️ **5.** Write your own problem. Write the number sentence to show how to solve it.

_____

_____

_____

_____

Problem Solving

Michelle has 3 more red marbles than blue ones. She has 25 marbles.

6. How many are blue?

_____ marbles are blue.

7. How many are red?

_____ marbles are red.

Chapter 14
Lesson 4

Making Subtraction Easier

NCTM Standards 1, 2, 6, 7, 8, 9, 10
🔻 TEKS 1.11A, 1.12A

Draw the jump.
Complete the number sentence.

1.

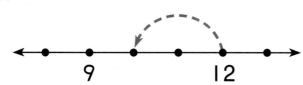

$$\boxed{12} - \boxed{2} = \boxed{10}$$

2.

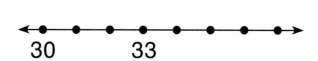

$$\boxed{36} - \boxed{} = \boxed{30}$$

3.

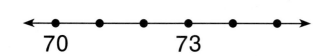

$$\boxed{} - \boxed{4} = \boxed{70}$$

4.

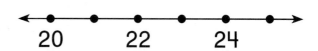

$$\boxed{25} - \boxed{5} = \boxed{}$$

5. Look at the number sentences in Problems 1 to 4.
How are they the same?

_____

_____

_____

_____

NOTE: Your child is learning different ways to make
subtraction easier. You might ask your child to explain
how he or she solved Problem 5.

Write the number sentence and answer the question.

6. Hector had 44¢. He spent a dime.
How much money did he have left?

_____¢

---

7. Katy's hair is 20 inches long.
Julie's hair is 9 inches long.

Who has longer hair? _____
How much longer is her hair?

_____ inches

Problem Solving

8. Champ, Spot, and Rover are dogs.
Champ weighs 3 pounds more than Spot.
Champ weighs 8 pounds less than Rover.
Champ weighs 17 pounds.
How much do the other dogs weigh?

Spot weighs _____ pounds.

Rover weighs _____ pounds.

Name _____ Date _____

Subtraction That Changes the Tens Digit

NCTM Standards 1, 2, 6, 7, 8, 9, 10

TEKS 1.11A

Draw the jump.
Complete the number sentence.

1.

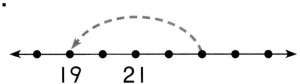

19   21

$\boxed{23} - \boxed{4} = \boxed{19}$

2.

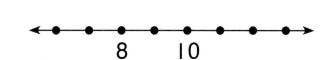

8   10

$\boxed{11} - \boxed{} = \boxed{8}$

3.

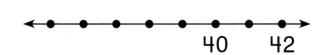

40   42

$\boxed{42} - \boxed{6} = \boxed{}$

4.

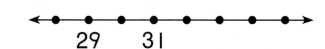

29   31

$\boxed{} - \boxed{5} = \boxed{29}$

5. Write the number sentence and answer the question.

Meg finished her homework in 43 minutes.
Kali was 8 minutes faster than Meg.
How long did Kali take?

$\boxed{} - \boxed{} = \boxed{}$ _____ minutes

NOTE: Your child is learning to subtract larger numbers in different ways. You might ask your child to explain how he or she solved Problem 5.

Write the number sentence and answer the question.

6. Tina's turtle weighs 22 ounces.

Tony's turtle weighs 5 ounces less.

How much does Tony's turtle weigh?   _____ ounces

7. It was 74°F outside in the morning.

It got 7°F colder by lunch time.

What was the temperature at lunch?   _____°F

$$\square - \square = \square$$

Problem Solving

8. Ann had 26 shells.

Ann gave half of them to Bill.

Then she gave 8 to Cindy.

How many shells do they each have now?

Bill has _____ shells.

Cindy has _____ shells.

Ann has _____ shells.

Name _____   Date _____

Change from a Dollar

NCTM Standards 1, 2, 6, 7, 8, 9, 10

TEKS 1.1C, 1.1D, 1.3A, 1.11A, 1.11D, 1.12A

Write the amount. Then write the change.

1. Jan spent this money.

_____5_____ ¢

She paid with a dollar bill.

She got _____95_____ ¢ change.

---

2. Don spent this money.

_____ ¢

He paid with a dollar bill.

He got _____ ¢ change.

---

3. Kelly spent this money.

_____ ¢

She paid with a dollar bill.

She got _____ ¢ change.

 NOTE: Your child is learning to make change from one dollar. You might ask your child how much change you get from a dollar when you buy a ball for 40¢.

Write the change. Then write the amount spent.
Use blocks to help.

4. This is my change from $1.

_____2̶1̶_____ ¢

I spent _____7̶9̶_____ ¢.

5. This is my change from $1.

_____ ¢

I spent _____ ¢.

6. This is my change from $1.

_____ ¢

I spent _____ ¢.

7. This is my change from $1.

_____ ¢

I spent _____ ¢.

8. This is my change from $1.

_____ ¢

I spent _____ ¢.

9. This is my change from $1.

_____ ¢

I spent _____ ¢.

Challenge

10. I have two coins.
How much money could I have?

_____

Name _____ Date _____

Is This Story Reasonable?

NCTM Standards 1, 2, 6, 7, 8, 9, 10
TEKS 1.11A, 1.12A, 1.13

Complete the story. Is it reasonable?
Circle *yes* or *no.*

1. Jack is twice as tall as his mother.
   His mother is 8 feet tall.

   Jack is ___16___ feet tall.

   Is this story reasonable?   yes   no

_____

2. It took Kelly 15 minutes to eat lunch.
   It took Young 10 minutes to eat lunch.

   _____ ate faster than _____.

   Is this story reasonable?   yes   no

_____

3. Munch and Chomp are monkeys.
   Munch ate 25 bananas in one week.
   Chomp ate 13 bananas that same week.

   Together the monkeys ate _____ bananas.

   _____ ate _____ more bananas than _____.

   Is this story reasonable?   yes   no

_____

NOTE: Your child is learning to decide whether a story is
likely or unlikely to happen. You may wish to ask your
child to tell a reasonable and an unreasonable story.

4. Write a reasonable story for 24 − 6.

_____

_____

_____

_____

5. Write an unreasonable story for 24 − 6.

_____

_____

_____

_____

Problem Solving

Complete the story. Is it reasonable?
Circle *yes* or *no*.

6. Larry's lizard ate _____ pounds of crickets,

_____ pounds of worms, and _____ pounds
of vanilla ice cream.

The lizard ate _____ pounds of food.

Is this story reasonable?   yes   no

Name _____   Date _____

Solving Puzzles with Many Solutions

NCTM Standards 1, 2, 6, 7, 8, 9, 10

Solve each puzzle.

1.

10		
	20	30
		60

2.

	3	43
30		
		78

3.

25	25	
13	12	

4.

93		93
7		
	18	

NOTE: Your child is learning to solve Cross Number Puzzles. Some puzzles have more than one solution. You might ask your child to find a different way to solve Problem 5.

5. Find two ways to solve the puzzle.

		50
		25
30	45	

		50
		25
30	45	

Challenge
What numbers are missing?

6.

		12
	14	
		26

7.

		0
15	20	

Write the missing numbers and signs. Lessons 1, 2

1.

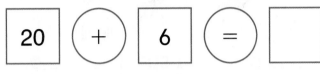

2.

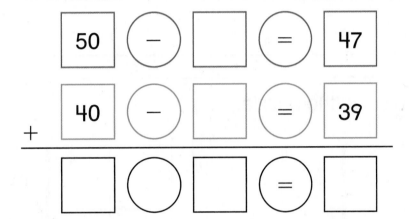

Write the number sentence and answer the question. Lessons 3, 4, 5

3. Avra walked 15 minutes in the morning.
She walked 12 minutes in the afternoon.

How many minutes did Avra walk?

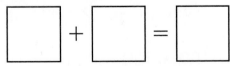

_____ minutes

4. Ben has 29¢. Ron has 12¢.
How much more money does
Ben have than Ron?

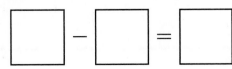

_____ ¢

5. Write the change. Then write the amount spent. Lesson 6
This is my change from $1.

_____¢   I spent _____¢.

---

6. Write a reasonable story for 20 + 24 = _____. Lesson 7

_____

_____

_____

7. Find two ways to solve the puzzle. Lesson 8

		60
		30
45	45	

		60
		30
45	45	

Problem Solving Lesson 9

8. Carlos uses toothpicks to make triangles.
He uses a different toothpick for each side.
How many toothpicks does he need to make
4 triangles?

_____ toothpicks

Chapter 15 # Exploring Rules and Patterns
What is the Rule? ✏️

STEP 1 **Completing the Table**

What numbers are missing?

in	1	2	3		5		7
out	3	4	5	6		8	

in

?

out

STEP 2 **Describing the Rule**

What is the rule for the machine? _____
Explain how you know.

_____

_____

_____

STEP 3 **Creating a New Rule**

Think of a rule for this machine.

_____

_____

in

?

out

Complete the table.

in	1	2	3	4	5		
out							

Investigation

Dear Family,

Today we started Chapter 15 in *Think Math!* In this chapter, I will use patterns to identify rules for rule machines. I will also use rules to explore how to convert different kinds of measurement units. There are NOTES on the Lesson Activity Book pages to explain what I am learning every day.

Here are some activities for us to do together at home. These activities will help me learn to recognize patterns and figure out rules to describe them.

Love,

_____

Family Fun

What's My Number?

Play this game with your child. Your child will also play this game in class.

- The first player picks a secret number smaller than 30.
- The second player tries to guess the number. For each guess, the first player responds with "too big," "too small," or "that's right" and records the number in a table like the one shown below.

Too big	Too small	That's it!
25	10	15
20	13	

- When the second player guesses the secret number, players switch roles and play a new game.

Making Rectangles

Work with your child to practice making rectangular arrays.

- You will need grid paper and a number cube.
- Take turns tossing the number cube two times to determine the number of rows and the number of columns in a rectangular array. Draw the array on the grid paper.

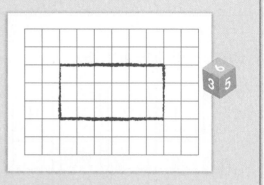

Name _____ Date _____

Identifying Rules

NCTM Standards 1, 2, 6, 7, 8, 9, 10
TEKS 1.10A, 1.12A

What is missing?
What is the rule?

1.

in	23	57	54	79		
out	20	54	51		63	37

The rule is _____.

2.

in	A	F	Q	W		
out	C	H	S		V	L

The rule is _____.

NOTE: Your child is learning to identify rules for rule machines by looking at inputs and outputs. Ask your child to explain how he or she found the rule for Problem 2.

What is missing?
What is the rule?

Use a calculator to help.

3.

in	40	12	31	27		
out	63	35	54		84	23

The rule is _____.

4.

in	clown	foot	rule	three	plus	
out	c	f				a

The rule is _____.

Challenge

5. Use the rule machine in Problem 4.
What words give **w** as an output?

four   add   flew

west   if   saw

then   swim   why   six   with

Name _____   Date _____

Sorting Rules

NCTM Standards 1, 2, 6, 7, 8, 9, 10
🔻 TEKS 1.4, 1.10A, 1.12A

What is missing?

1.

in	2	53	31	99	20	
out	yes	no	no			no

2.

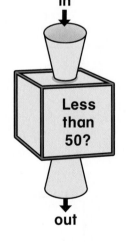

in	31	78	3	49	53	
out	yes	no	yes			yes

NOTE: Your child is learning that rules can be used for sorting. Ask your child to explain how the machine in Problem 1 sorts numbers.

What is missing?

3.

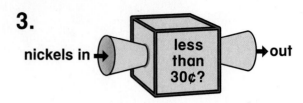

nickels in → [less than 30¢?] → out

in	out
1	yes
2	yes
6	no
3	
7	
	no

4.

inches in → [longer than 1 foot?] → out

in	out
6 in.	no
14 in.	yes
9 in.	no
11 in.	
24 in.	
	yes

12 inches = 1 foot

Challenge

5. What is missing?
What is the rule?

in ↓
?
out ↓

in	5	2	10	4	
out	11	5			7

The rule is _____.

Name _____ Date _____

Undoing Rules

NCTM Standards 1, 2, 6, 7, 8, 9, 10

TEKS 1.4, 1.10A, 1.12A

Write the missing numbers.
What is the rule?

1.

in

out

in	20	15	40	33		
out	30	25	50		60	13

The rule is _____.

2.

in

out

in	67	40	25	80		
out	57	30			50	31

The rule is _____.

3. What do you notice about the rules for
Problems 1 and 2?

_____

_____

NOTE: Your child is exploring rules that undo each other.
Ask your child to explain how the rules in Problems 1 and 2
are alike and how they are different.

Write the missing numbers.
Use a calculator if you like.
What is the rule?

4.

in	2	10	9			12
out	20	28		38	18	

The rule is _____.

5.

in	12:00	1:30	4:00	:	:	7:00
out	12:30	2:00	:	1:30	5:00	:

The rule is _____.

Challenge

6. Write rules to undo the rules in Problem 4 and Problem 5.

Problem 4 _____

Problem 5 _____

Chapter 15
Lesson 4

Rules with More Than One Input

NCTM Standards 1, 2, 6, 7, 8, 9, 10
TEKS 1.4, 1.10A, 1.12A

What is missing?

1.

in	6	10	15	20	17	48
in	2	8	5	15		
out	4	2	10		13	18

2.

D	1	2	5	3	4	
P	1	3	0	9	8	
¢	11¢	23¢	50¢	¢	¢	99¢

NOTE: Your child is learning to apply
rules with more than one input. Ask your
child to explain the rule in Problem 2.

Write the missing numbers.

3.

in	1	5		33	67	
in	3	7	51	35	69	
out	2	6				55

4.

in	5	14	79	40	25	10
in	8	23	76	12	15	6
in	20	6	70	18	5	
out	20	23	79			10

Challenge

5. If I know an output for Problem 3, then I know both inputs because

_____

_____

Conversion Rules

NCTM Standards 1, 2, 3, 4, 6, 7, 8, 9, 10

TEKS 1.10A, 1.12A

Write the missing numbers.

1.

Number of pints	1	2	3	4	7	
Number of cups	2	4	6			20

2.

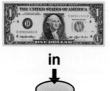

$1	1	2	3	4		10
D	10	20	30		80	

NOTE: Your child is learning to convert from one unit to another. Ask your child to explain how he or she solved each of the problems.

Write the missing numbers.
Use a calculator if you like.

3.

Number of quarters	1	2	3	6	10	4	
Number of nickels	5	10	15				25

4.

Number of gallons	1	3	10	2	5	6	
Number of quarts	4	12	40				16

Challenge

5.

Number of yards	1	2	3	5	7	
Number of feet	3	6	9			18

6. What is the rule? _____

7. Explain how you found the missing numbers.

_____

_____

Name _____ Date _____

Skip-Counting with Money

NCTM Standards 1, 2, 4, 6, 7, 8, 9, 10
TEKS 1.1C, 1.5D, 1.11A, 1.12A

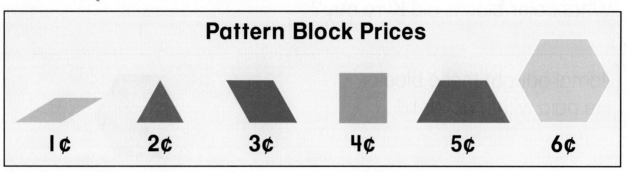

Pattern Block Prices

1¢   2¢   3¢   4¢   5¢   6¢

Use the prices above.
How much will the blocks cost?

1. 2 □

 ____ ¢

2. 4 ⬱ ____ ¢

3. 6  ____ ¢

4. 3 ____ ¢

5. 5 ____ ¢

6. 4 ⬡ ____ ¢

7. 3 □ and 2 ▲ ____ ¢

8. 2 ⬱ and 1 ____ ¢

 NOTE: Your child is skip-counting to find the prices for groups of pattern blocks. Ask your child to explain how he or she found the value of the blocks in Problem 7.

9. Kyra bought 4 blocks.
They cost 20¢.
What color blocks did Kyra buy? _____

---

10. Jamal bought these blocks.
He paid with a dollar bill.

How much did the blocks cost? _____¢

How much was his change? _____¢

---

11. Sue bought twice as many  as 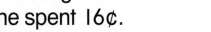 .
She spent 16¢.
What did she buy?

_____ and _____

---

12. Dex bought one kind of block.
He paid 18¢.
What color could his blocks be?

_____ or _____

Problem Solving

13. Tamara spent 12¢. She got 3 blocks.
They were **not** all the same.
What color could her blocks be?

_____, _____, and _____

Creating Figures and Patterns

NCTM Standards 1, 2, 6, 7, 8, 9, 10

TEKS 1.9B, 1.10A, 1.12A

Complete each table.

1. I am making fish.

Number of fish	1	2	3	4	5	6
Cost of ▱	3¢	6¢	9¢			
Cost of ▲	2¢	4¢	6¢			
Total cost	5¢	10¢				

2. I am making computers.

Number of computers	1	2	3	4	5	6
Cost of ■	4¢					
Cost of ▱	1¢					
Total cost	5¢					

NOTE: Your child is learning to look for patterns to help find sums. Ask your child to describe a pattern in Problem 1.

Complete each table.

3. I am making fancy hexagons.

Number of hexagons	1	2	3	4	5	6
Cost of ⬢ (trapezoid)						30¢
Cost of ⬢ (parallelogram)				12¢		
Cost of ▲ (triangle)			6¢			
Total cost						

---

4. I am making houses.

Number of houses	1	2	3	5	7	10
Cost of (trapezoid)	5¢					
Cost of (square)	4¢					
Total cost	9¢					

320 three hundred twenty **CCCXX** 160 + 160

Name _____ Date _____

Patterns with Skip-Counting

NCTM Standards 1, 2, 6, 7, 8, 9, 10

1. Skip-count on the grid below.
   Mark jumps of 4 and 6.

Mark jumps
of 4 with an X.

Mark jumps
of 6 with an ○.

1	2	3	4	5	6	7	8	9	10
11	12	13	14	15	16	17	18	19	20
21	22	23	24	25	26	27	28	29	30
31	32	33	34	35	36	37	38	39	40

NOTE: Your child is looking for patterns
while skip-counting. Ask your child to describe
any patterns in the grid above.

2. Skip-count on the grid below.
   Mark jumps of 8 with an ✕.
   Mark jumps of 7 with an ◯.

1	2	3	4	5	6	7	8	9	10
11	12	13	14	15	16	17	18	19	20
21	22	23	24	25	26	27	28	29	30
31	32	33	34	35	36	37	38	39	40
41	42	43	44	45	46	47	48	49	50
51	52	53	54	55	56	57	58	59	60

---

3. Look at the grid in Problem 2.
   Where do the jumps meet?
   Find ⊠.

   _____

Problem Solving

4. What numbers are missing?
   I start at 0.

   I make _____ jumps of 6 or _____ jumps of 7

   to get to _____.

Chapter 15
Lesson 9

Relating One-Color Trains

NCTM Standards 1, 2, 6, 7, 8, 9, 10

TEKS 1.4, 1.12A

Write the missing numbers.

1.

_____ is as long as _____ .

2.

_____ is as long as _____ .

3.

_____ are as long as _____ .

4.

_____ are as long as _____ .

5.

_____ are as long as _____ .

NOTE: Your child is exploring multiples by building Cuisenaire® Rod trains using the same color blocks. Ask your child to explain how he or she solved Problem 3.

Write the missing numbers.

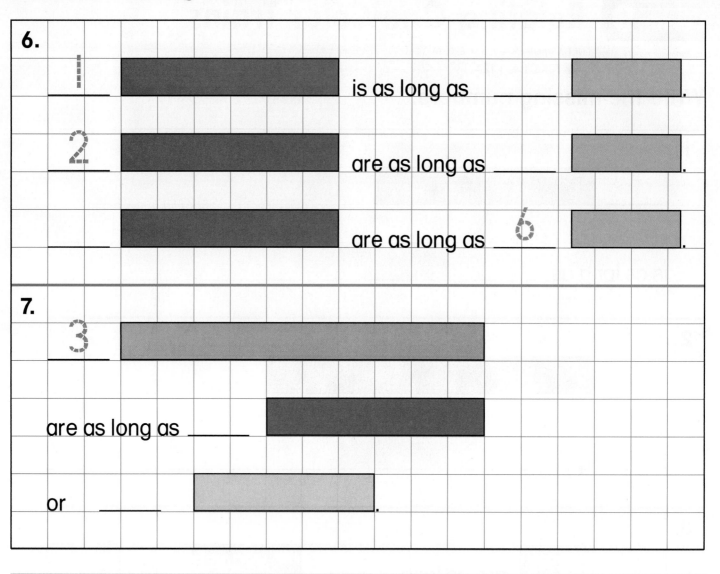

6.

___1___ [bar] is as long as ___ [bar].

___2___ [bar] are as long as ___ [bar].

___ [bar] are as long as ___6___ [bar].

7.

___3___ [bar]

are as long as ___ [bar]

or ___ [bar].

✎ **8.** What pattern do you notice in Problem 6?

_____

_____

Challenge

9. Build matching trains. Write the missing numbers.

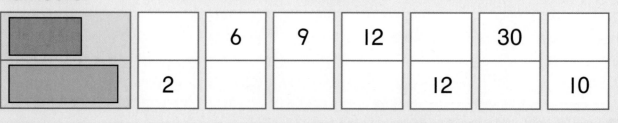

		6	9	12		30	
	2				12		10

Problem Solving Strategy
Work Backward

NCTM Standards 1, 2, 6, 7, 8, 9, 10

TEKS 1.11A, 1.11B, 1.11C, 1.12A

Understand

Plan

Solve

Check

1. Joe had some money.
   He got 27¢ more.
   He spent 15¢.
   Now he has 27¢.
   How much money did Joe start with?

   _____¢

2. Wilbur collects stuffed toy bears.
   He got 2 new bears every year.
   When he was 10 years old, he had 20 bears.
   How many bears did he have
   when he was 4 years old?

   _____ bears

3. Paula makes quilts.
   Each year she makes 1 less quilt
   than she did the previous year.
   This year she made 3 quilts.
   How many quilts did she make 4 years ago?

   _____ quilts

NOTE: Your child is using the strategy, *work backward,* to solve
problems. Ask your child to explain how he or she
solved the problems on this page.

Problem Solving Test Prep

1. Lin puts 12 red and white
   flowers in a vase.
   There are 4 more red flowers
   than white flowers.
   How many red flowers
   are there?

   (A) 4   (C) 8

   (B) 6   (D) 12

2. Ethan makes items for the craft
   fair. He makes 3 bookmarks
   and 4 cards each day.
   How many items will he make
   in 3 days?

   (A) 3   (C) 7

   (B) 4   (D) 21

 Show What You Know

3. Hannah has 12 square tiles.
   How many different rectangles
   can she make?

   _____ rectangles

   Use words, numbers,
   or pictures to explain.

4. Kate skip-counts by twos.
   Dan skip-counts by threes.
   They both start at 0.
   What is the first number both
   Kate and Dan will say?

   _____

   Explain how you found the
   answer.

   _____

   _____

   _____

   _____

   _____

Name _____ Date _____

Review/Assessment

NCTM Standards 1, 2, 3, 4, 6, 7, 8, 9, 10

Write the missing numbers.
What is the rule? Lessons 1, 2, 3, and 4

1.

in

?

out

in	4	12	10		2	21
out	8	16	14	5		

The rule is _____.

Write the missing numbers. Lesson 5

2.

Number of quarts	1	2	3	5		7	
Number of cups	4	8	12		40		16

What is the cost? Lessons 6 and 7

3. One of these costs _____¢.

4. Three of these cost _____¢.

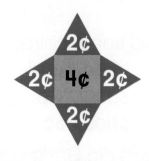

© Education Development Center, Inc.

Skip-count on the grid below. Lesson 8

5. Mark jumps of 3 with an ✕.
   Mark jumps of 4 with an ○.

1	2	3	4	5	6	7	8	9	10
11	12	13	14	15	16	17	18	19	20
21	22	23	24	25	26	27	28	29	30

6. Where do the jumps above meet? Lesson 8

   Find ⊠.

   _____

Write the missing number. Lesson 9

7. _____ are as long as

   2 _____.

Problem Solving Lesson 10

8. Carla had some coins.
   She found 23¢ more.
   Then she spent 12¢.
   Now Carla has 26¢.
   How much did Carla start with? _____¢

add

$3 + 2 = 5$

addend

$$\mathbf{5} + \mathbf{2} = 7$$

addend   addend

addition sentence

$4 + 1 = 5$

after

18 19 20

20 is **after** 19.

afternoon

always

Green will **always** be picked.

area

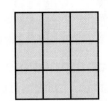

area = 9 square units

array

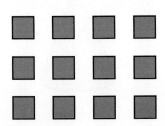

attribute

small, red circle   large blue circle

backward

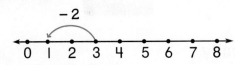

-2

0 1 2 3 4 5 6 7 8

bar graph

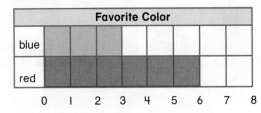

Favorite Color								
blue								
red								

0 1 2 3 4 5 6 7 8

base-ten blocks

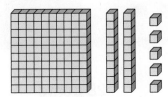

before

18 19 20

18 is **before** 19.

blue

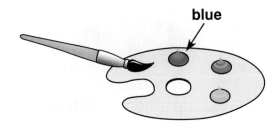

blue

bottom

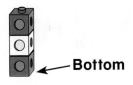

Bottom

calculator

capacity

cent (¢)

28¢

28 cents

centimeter

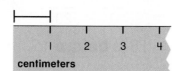

certain

Picking green is **certain**.

change

circle

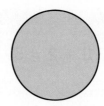

clock

closed figure

column

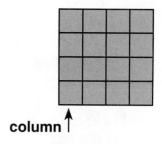

column ↑

compare

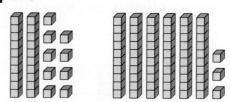

When you **compare** the two groups, you see that 29 is less than 63.

cone

congruent

corner

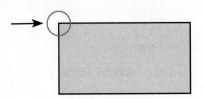

cube

cylinder

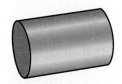

data

Kinds of Books We Like	
Funny	▢▢▢▢▢▢▢▢▢
Fantasy	▢▢▢▢▢
Sports	▢▢
Real-Life	▢▢▢
Biography	▢

Key: Each ▢ stands for 1 child's choice.

day

The **days** of the week are:
**Sunday, Monday, Tuesday,
Wednesday, Thursday, Friday,**
and **Saturday.**

degree symbol (°)

72° ◄— **degree symbol**

difference

$$9 - 3 = 6$$

← **difference**

different

different shape and color

digit

51

digit ↑ ↑ **digit**

dime

 or **10¢
10 cents**

dollar

1 dollar = 100¢

dollar sign

$

double

$$4 + 4 = 8$$

down

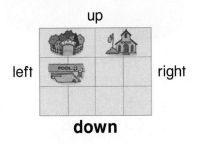

up

left   right

down

east

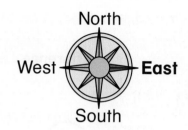

equal

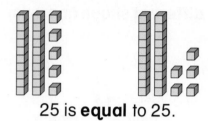

25 is **equal** to 25.

equal sign (=)

$$4 + 1 = 5$$
↑
equal sign

estimate

about 10 buttons

even numbers

0, 2, 4, 6, 8, 10 . . .

evening

fact family

5 + 3 = 8   3 + 5 = 8
8 − 3 = 5   8 − 5 = 3

feet

Use **feet** to measure longer objects.

fewer

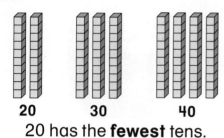

20   30

20 has **fewer** tens than 30.

fewest

20   30   40

20 has the **fewest** tens.

fifty

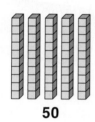

50

first

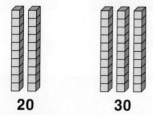

first   second   third   fourth   fifth

forty

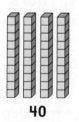

40

forward

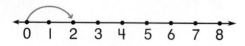

fourth

first   second   third   fourth   fifth

fraction

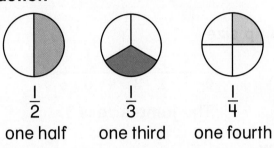

$\dfrac{1}{2}$ one half $\dfrac{1}{3}$ one third $\dfrac{1}{4}$ one fourth

gallon

grid

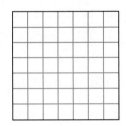

half

Half of the cubes are red.

half hour

4:30

There are 30 minutes in a **half hour.**

halfway

3 is **halfway** between 1 and 5.

heavier, heaviest

The rock is the **heaviest** object.
The rock is **heavier** than the feather.

height

height

hexagon

horizontal

This line is **horizontal.**

hour

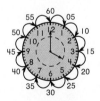

4:00

There are 60 minutes in an **hour.**

hour hand

hour hand

hundreds

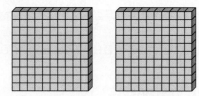

impossible

Picking red is **impossible.**

inch(es)

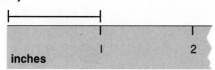

input

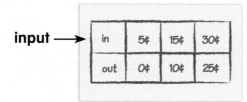

intersection

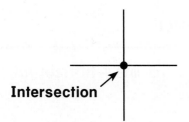

Intersection

inverse

$5 + 4 = 9$, so $9 - 4 = 5$.

is equal to (=)

25 **is equal to** 25.

$25 = 25$

is greater than (>)

5 **is greater than** 1.

$5 > 1$

is less than (<)

3 **is less than** 5.

$3 < 5$

jump

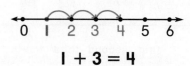

$1 + 3 = 4$

There are 3 **jumps** between 1 and 4.

jump size

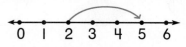

The **jump size** is 3.

key

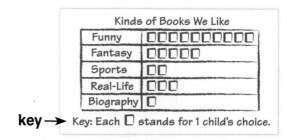

key → Key: Each ☐ stands for 1 child's choice.

landing number

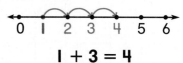

$1 + 3 = 4$

The **landing number** is 4.

large

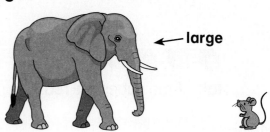

← large

largest

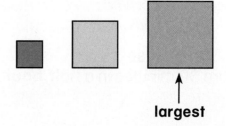

largest

last

first   next   last

left

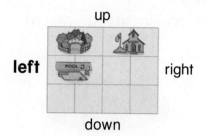

length

The spoon has a **length** of 6 inches.

lighter, lightest

A pencil is **lighter** than a book.
The pencil is the **lightest** object.

likely

It is **likely** that I will pick blue.

line

line of symmetry

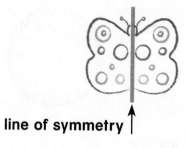

line of symmetry ↑

liter

longer

long

longer →

longest

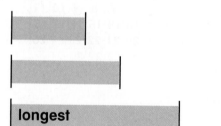

longest

middle

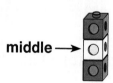

middle →

minus (−)

$$3 - 2 = 1$$
3 **minus** 2 equals 1.

minute hand

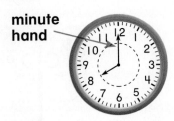

minute hand

mirror image

Both parts match.

missing addend

$$5 + \boxed{} = 8$$

↑
missing addend

month

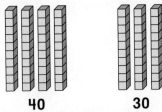

more

40   30

40 has **more** tens than 30.

morning

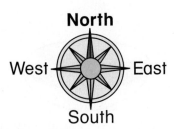

most

The jar with the red lid has the **most** coins.

multiples of ten

10, 20, 30, 40, 50, 60, 70 . . .

never

Blue can **never** be picked.

next

first   next   last

nickel

 or   5¢
5 cents

north

North

West — East

South

number

A **number** has one or more digits.

number line

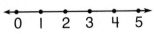

0 1 2 3 4 5

number sentence

$$4 + 2 = 6$$
or
$$6 - 3 = 3$$

o'clock

The clock shows 1 **o'clock.**

odd numbers

1, 3, 5, 7, 9 . . .

one hundred

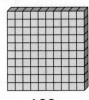

100

one, two, three . . .

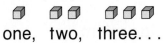

one, two, three. . .

ones

There are 3 **ones.**

open figure

order

0   30   58   61   85   100

These numbers are in **order** from smallest to biggest.

ounce (oz)

A slice of bread weighs about one **ounce.**

output

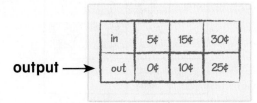

output →

parallelogram

path

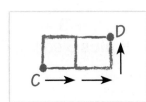

pattern

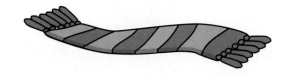

pattern unit

penny (pennies)

 or   1¢
1 cent

pentagon

picture graph

pint

plus (+)

$$4 + 3 = 7$$
4 **plus** 3 equals 7.

possible

It is **possible** to pick a blue marble.

pound (lb)

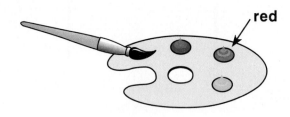

A loaf of bread weighs
about one **pound.**

price

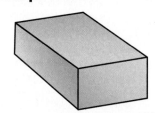

quart

quarter

 or   25¢
25 cents

reasonable

A girl eats 2 apples. ← **reasonable**
A girl eats 20 apples. ← unreasonable

rectangle

rectangular prism

red

red

rhombus

right

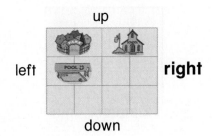

up · left · right · down

rods

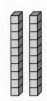

round

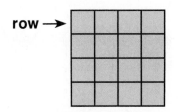

row

row →

rule

Add 2	
4	6
6	8
8	10

ruler

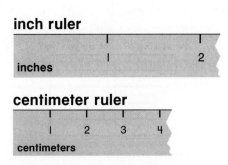

inch ruler — inches
centimeter ruler — centimeters

same

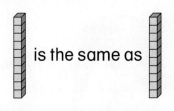

is the same as

second

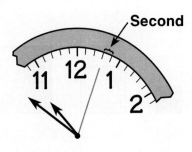

Second
11 12 1 2

shorter

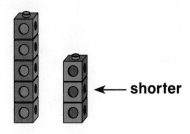

← shorter

shortest

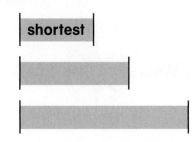

shortest

side

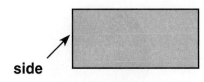

side

skip-count

2, 4, 6, 8, 10
Skip-count by twos.

slide

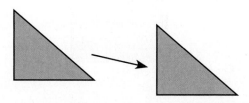

small

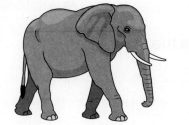

small

smallest

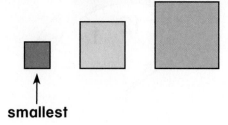

smallest

sort

triangles   not triangles

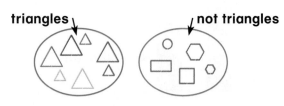

south

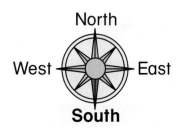

North

West   East

South

sphere

square

square units

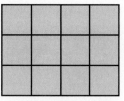

12 square units

Stair-Step Numbers

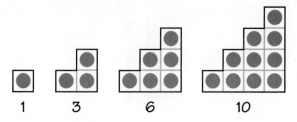

1   3   6   10

starting number

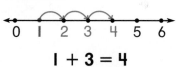

0 1 2 3 4 5 6

$1 + 3 = 4$

The **starting number** is 1.

subtract

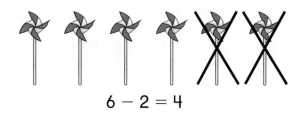

$6 - 2 = 4$

subtraction sentence

$8 - 2 = 6$

sum

$5 + 1 = 6$

sum

survey

What kind of book do you like?	
Funny	||||‖ ||||‖
Fantasy	||||‖
Sports	||
Real-Life	|||
Biography		

take away

5 **take away** 2 is 3.

tally

 ← **tally mark**

temperature

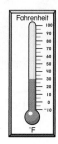

The **temperature** is 65 degrees.

tens

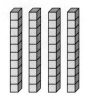

thermometer

third

first   second   third   fourth   fifth

thirty

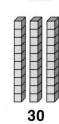

30

today

The day that is right now.

tomorrow

The day that is after today.

top

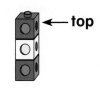 ← **top**

trapezoid

triangle

turn

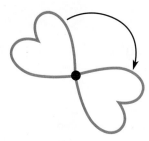

twenty

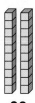

20

undo

$5 - 3 = 2$
$2 + 3 = 5$
Addition is used to **undo** subtraction.

units

5 **units**

unlikely

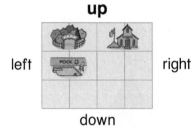

It is **unlikely** the spinner will land
on yellow.

unreasonable

A boy jumps 1 foot. ← reasonable
A boy jumps 60 feet. ← **unreasonable**

up

up

left   right
down

value

The **value** of the coins is 13¢.

vertical

This tower is **vertical.**

week

week →

weigh

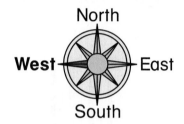

A slice of bread **weighs**
about 1 oz.

west

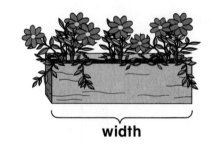

width

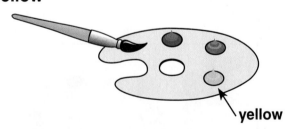
width

yellow

yellow

yesterday
The day before today.